FUNDEMENTALS FOR A BOTANY LAB MANUAL

Thomas Smith, Ph.D.
Ave Maria University

Cover image © 2015

www.kendallhunt.com
Send all inquiries to:
4050 Westmark Drive
Dubuque, IA 52004-1840

ISBN 978-1-4652-6620-0

FUNDEMENTALS OF BOTANY LABORATORY

TABLE OF CONTENTS

PREFACE

TO THE STUDENT

THIS COURSE IS ONE TAKEN BY BIOLOGY MAJORS/MINORS OR STUDENTS IN OTHER AREAS WHO'S CURRICULUM REQUIRES THE COURSE. OBVIOUSLY, THIS SUGGESTS THAT EACH OF YOU IS A STUDENT OF BIOLOGY. AS A RESULT, THOSE OF US INVOLVED IN TEACHING THE COURSE ASSUME THAT EACH OF YOU WILL DEMONSTRATE **INTEREST** IN AND **INQUISITIVENESS ABOUT** THE MATERIALS PRESENTED AND WILL BE **CONSCIENTIOUS ABOUT** THE STUDY OF THE MATERIALS. IF THIS ASSUMPTION IS INCORRECT IN RESPECT TO YOU PERSONALLY, IT WOULD BE ADVISABLE FOR YOU TO DROP THE COURSE. THE LAB IS TO A GREAT EXTENT SELF-PACED EXPERIENCE; YOUR LAB INSTRUCTOR IS PRESENT TO ASSIST YOU IN YOUR STUDIES AND TO ANSWER QUESTIONS THAT YOU MAY HAVE.

In order to save yourself time and to get the most of each laboratory experience, it is important that you read the laboratory exercises before you come to each laboratory meeting.

Bring your textbook each laboratory meeting as you will want to refer to figures and illustrations in the textbook.

PLEASE NOTE: Bringing **food** or **drink** into or using **tobacco** products in the laboratory is not permitted! Additionally, please remember that if you make a mess, you clean it!! GOOD LUCK!!

A WORD ON BIOLOGICAL CLASSIFICATION AND PLANTS

Finding Order in Diversity

Humans do not attempt to give each individual object in nature a unique name. Instead, we classify items; we aggregate things into groups or sets – e.g., "chairs, boats, bryophytes, euglenoids, people," etc. -- the members of which have certain common features. In biology, that particular endeavor of naming and classifying organisms is called **TAXONOMY**.

Why do we classify? We simply could not think or communicate rationally and effectively if each individual's object in the world had a unique name! Among all the people on Earth, think for a moment about how many of them you known by their individual name!

As previously suggested, classification involves the grouping together of individual things into sets. This is done by observing their characteristics; all individuals having attributes in common are placed into the same set since they are similar to a greater or lesser degree. Once the sets are constructed, each is designated by a name; an important function of the name is **symbolic communication**.

By definition, the name of a set may furnish information about, for example, color, shape, size and use of the object, as well as its relationship to other objects. For example, all objects having a large, flat, horizontal surface supported by four vertical members may be grouped together into a set; this particular set may be named, "table". Using this method, other sets of objects may be distinguished and named; for example, a set of "chairs" and a set of "beds". An even more inclusive set of objects may be recognized and named, i.e., a set named, "furniture". "Tables", "chairs" and "beds" become subsets within the larger, more inclusive set, "furniture". Such classification by using sets and subsets may be continued through a large number of levels. The result is a **hierarchy** sequence of sets at different levels in which each set except the lowest includes one or more subsets

Inanimate Objects
 Human-Made Objects
  Household Goods
   Furniture
    Tables (or Beds, Chairs, etc.)

In the classification of living organisms, a particular form of hierarchy is used with each level (= **taxon**; pl., taxa) in the hierarchy:

Kingdom
 Division (Phylum in Zoological Nomenclature)
  Class
   Order
    Family
     Genus (pl., genera)
      Species

Each level in the hierarchy includes a greater variety of organisms than the level directly beneath it. The most inclusive taxon is the Kingdom, the least inclusive is the species (or, in some cases, a subunit of the species). An organism is completely classified when it has been assigned to some definite set at each of the seven levels. A full classification of a single species, the dinoflagellate, which is the causative agent of "Red Tide", would be:

Kingdom Protista
 Division Dinophyta
  Class Dinophyceae
   Order Gymnodiniales
    Family Gymnodiniaceae
     Gymnodinium brevis

The generic name and the specific epithet comprise what is commonly called the **scientific name**; more formally, it is referred to as the **binomial**, meaning literally, "two names". The modern approach of naming organisms and placing them into a system of classification is not left to the whims of the biologist personally involved. There are rules in the **International Code of Botanical Nomenclature**. One of the most basic rules is that each species may have only **valid binomial**. You can imagine the confusion that would result were there not such a rule!

The same scientific names are used throughout the world. This uniformity of usage ensures that each scientist will know exactly which species another scientist is citing. There would be no such assurance if common names were used; not only would any given species have a different name in each language, but often it would have more than one name in the same language! Thus, the fundamental value of **symbolic communication** would be lost.

One must recognize that any system of classification devised by humans is subject to an individual's interpretation of nature and, perhaps, to a degree of personal bias. Additionally, as our knowledge increases over time, systems of classification are subject to revision. Each system represents an attempt to portray orderliness the biologist assumes to be present in nature; it is hoped that the orderliness is not merely an artifact. Systems of classification **ideally** reflect nature interrelationships among organisms. The extent to which a given system of classification accomplishes this objective is often a matter of opinion. As a student, you should avoid accepting a particular system as dogma. On the contrary, you should be critical and aware of their ephemeral nature. What is generally accepted today as a reasonable representation of natural relationships may change tomorrow in light of new information? There are many systems of classification; there is no such thing as **system of classification**!

The diversity among living organisms is enormous. Faced with such great diversity, biologists have devised systems of classification, attempting to arrange the organisms into scientifically logical groupings to aid in their study.

In the 19th and early 20th centuries all organisms, excluding the viruses, were treated by many biologists as being either plants (**Kingdom Plantae**) or animals (**Kingdom Animalia**). However, even during that period of time, it was not uncommon to encounter in some systems of classification a third group, the **Protista**, in which were placed essentially unicellular organisms, many flagellated, -- some heterotrophic, others photoautotrophic -- which did not fit well the traditional concept of all organisms being either plant or animal. Depending upon one's personal bias, these could be considered to be plants with animal-like affinities, i.e., "**planimals**", or animals with plant-like affinities, i.e., "aniplans"! In any case, a common system of classification used for the plant kingdom through the middle of the 20th century was:

The acceptance of the **Kingdom Protista** is widely accepted today.

Please keep in mind that placing two or more organisms in the same taxon within a **system of classification suggests that they are naturally related to one another**! With that in mind it is clear that, among oxygenic, photoautotrophic organisms, which include some prokaryotic bacteria, the prokaryotic cyanobacteria and chloroxybacteria, and the various groups of eukaryotic algae and the plants, all are **ancestrally** related!

The fundamental differences in organization between prokaryotic and eukaryotic cells was recognized by some biologists in the mid-19th century. However, it was not until the advent of transmission electron microscopy that the extent and detail of those differences were revealed. Today, most systems of classification recognize that fundamental difference by segregating the prokaryotic organisms into the **Kingdom Prokaryota (=Kingdom Monera**, of some authors). **The bacteria, cyanobacteria and chloroxybacteria are not, in the modern sense of the word, plants!**

The oxygenic algal groups, excluding the prokaryotic cyanobacteria and chloroxybacteria, are believed to represent several evolutionary series only distantly, if at all, related to one another. Furthermore, though all have oxygen-evolving photosynthetic systems utilizing chlorophyll a as the primary photoreceptive pigment, none has achieved tissue/organ level organization, nor do they possess the adaptations for life in the terrestrial environment. They are placed into the **Kingdom Protista**. **The various eukaryotic algal groups are not, in the modern sense of the word, plants!**

Thus, our modern concept of the **Kingdom Plantae** is such that it is considered to include only those oxygenic, photoautotrophic organisms which attain tissue/organ level organization, have a distinct cytological and morphological alternation of generations, i.e., are diplobiontic and heteromorphic, and are embryophytic. This course is specifically concerned with the biology of oxygenic, photoautotrophic organisms.

The system of classification is **largely** that adopted for your textbook for extant groups only.

PROKARYOTIC ORGANISMS

KINGDOM PROKARYOTA (= MONERA)
SUBKINGDOM EUBACTERIA
Division Cyanobacteria
Class Cyanophyceae -- cyanobacteria (= "blue-green algae")
Division Prochlorophyta (= Chloroxybacteria)
Class Prochlorophyceae -- prochlorophytes (="chloroxybacteria")

EUKARYOTIC ORGANISMS

KINGDOM PROTISTA

Division Chrysophyta
Class Chrysophyceae -- "golden brown algae"

Division Xanthophyta
  Class Xanthophyceae -- "yellow green algae"
Division Bacillariophyta
  Class Bacillariophyceae -- "diatoms"
Division Dinophyta
  Class Dinophyceae -- "dinoflagellates"
Division Cryptophyta
  Class Cryptophyceae -- "cryptomonads"
Division Phaeophyta
  Class Phaeophyceae -- "brown algae"
Division Rhodophyta
  Class Rhodophyceae -- "red algae"
Division Euglenophyta
  Class Euglenophyceae -- "euglenoids"
Division Chlorophyta
  Class Chlorophyceae
   sensu stricto -- chlorophytes (= "chlorophycean green algae")
  Class Charophyceae
   sensu lato -- charophytes (= "charophycean green algae")

KINGDOM PLANTAE

"BRYOPHYTES"
Division Hepatophyta -- "liverworts"
Division Anthocerophyta -- "hornworts"
Division Bryophyta -- "mosses"

CRYPTOGAMOUS VASCULAR PLANTS"
Division Psilotophyta -- "psilopsids"
Division Lycophyta -- "lycopsids"
Division Sphenophyta -- "sphenopsids"
Division Pterophyta -- "pteropsids"

"PHANEROGAMOUS VASCULAR PLANTS"
"GYMNOSPERMS"

Division Cycadophyta -- "cycads
Division Ginkgophyta -- "ginkgo"
Division Gneophyta -- "gnetophytes"
Division Coniferophyta -- "conifers"

"ANGIOSPERMS"

Division Anthophyta
  Class Dicotyledoneae -- "dicots"
  Class Monocotyledoneae – "monocots"

LABORATORY 1.
CARE AND USE OF THE COMPOUND LIGHT MICROSCOPE

Since biological objects may be very small, it is often necessary to use a magnifying system to study them. We have available many devices serving effectively as magnifiers from simple hand lenses to electron microscopes. During this course you will routinely use a binocular compound light microscope. It is critical that you learn to use the instrument properly and effectively.

On some occasions, you will need to use a dissecting microscope. Your instructor will assign a **specific** OLYMPUS CHT-Series binocular compound microscope to you. "Microscope will bear a number identical to the seat you occupy in the laboratory". You are to use and care for this instrument during the course of the semester. If, during its use, you should encounter a problem, it is important to call the problem to the attention of the lab instructor immediately. After the lab instructor has demonstrated the proper technique for carrying the microscope, take "microscope from the shelf and place it at your work station. The lab instructor will introduce you to the use of the microscope. Follow her/his instructions explicitly! Identify the following parts of the microscope assigned to you.

ARM -- Supports and stabilizes microscope parts; serves as carrying handle.

BINOCULAR EYEPIECE/OCULAR LENSES -- Topmost series of lenses through which an object is viewed. The microscope has 10X ocular lenses, one of which has a pointer in it. The left-hand ocular lens has diopter adjustment.

BINOCULAR LENS (BODY) TUBE -- Holds revolving nosepiece and conducts light to objective lenses.

REVOLVING NOSEPIECE -- Holds objective lenses. As the nosepiece is turned, a different objective lens is rotated into position above the stage of the microscope.

OBJECTIVE LENSES:

4X (Scanning) Objective -- This lens has the largest field of view and the greatest working distance; bears a **red** ring identifier.

10X (Low Power) Objective -- Used to view objects in greater detail than possible with the scanning objective; bears a **yellow** ring identifier.

40X (High Power) Objective -- Used to view objects in greater detail than possible with either the scanning or low power objective; bears a **blue** ring identifier.

MECHANICAL STAGE -- Surface upon which a microslide is positioned for observation. The stage has coaxial control knobs on the bottom, rear, of the stage by means of which the microslide can be moved right-left and up down.

FOCUSING KNOBS -- Coaxial controls at the bottom, rear, of the microscope base by means of which an object is brought into focus. One, the larger, is for "coarse adjustment" to bring an object into approximate focus; the other, smaller knob if for "fine adjustment" to bring an object into final focus. **DO NOT USE THE COARSE ADJUSTMENT FOR ANY LENS EXCEPT THE 4X AND 10X OBJECTIVES!!!!!! Failure to follow this directive may result in breakage of the microslide and/or damage to the objective lens!** You must also be careful not to turn the focusing knobs past their upper or lower limits. BE GENTLE!

SUBSTAGE CONDENSER -- Adjustable (up/down) with a small knob on the right-hand side beneath the mechanical stage. Adjustment facilitates control of light in respect to the objective lens in use. Be careful not to turn the adjustment knob pas its upper or lower limits. BE GENTLE!

IRIS DIAPHRAGM -- Located on the front side of the substage condenser, this small lever opens/closes to control the amount of light passing through the condenser and into the objective lens. The overlapping "leaves" of the diaphragm are delicate and easily damaged. BE GENTLE!

ILLUMINATOR -- Provides source of light for objective lens. On/off switch is on the right-hand, front of the microscope base; you can control the intensity of the light by rotating the knob on the left-hand, back of the microscope base.

Using your microscope, perform the following. **If you do not understand any step in the instructions, STOP and request the assistance of your lab instructor!**

1. Rotate the nosepiece until the 4X objective is in place above the mechanical stage.
2. With the coarse adjustment knob, lower the microscope stage until it stops. Do not force the adjustment knob past the "stop point!"

Figure 1.1 Microscope Diagram

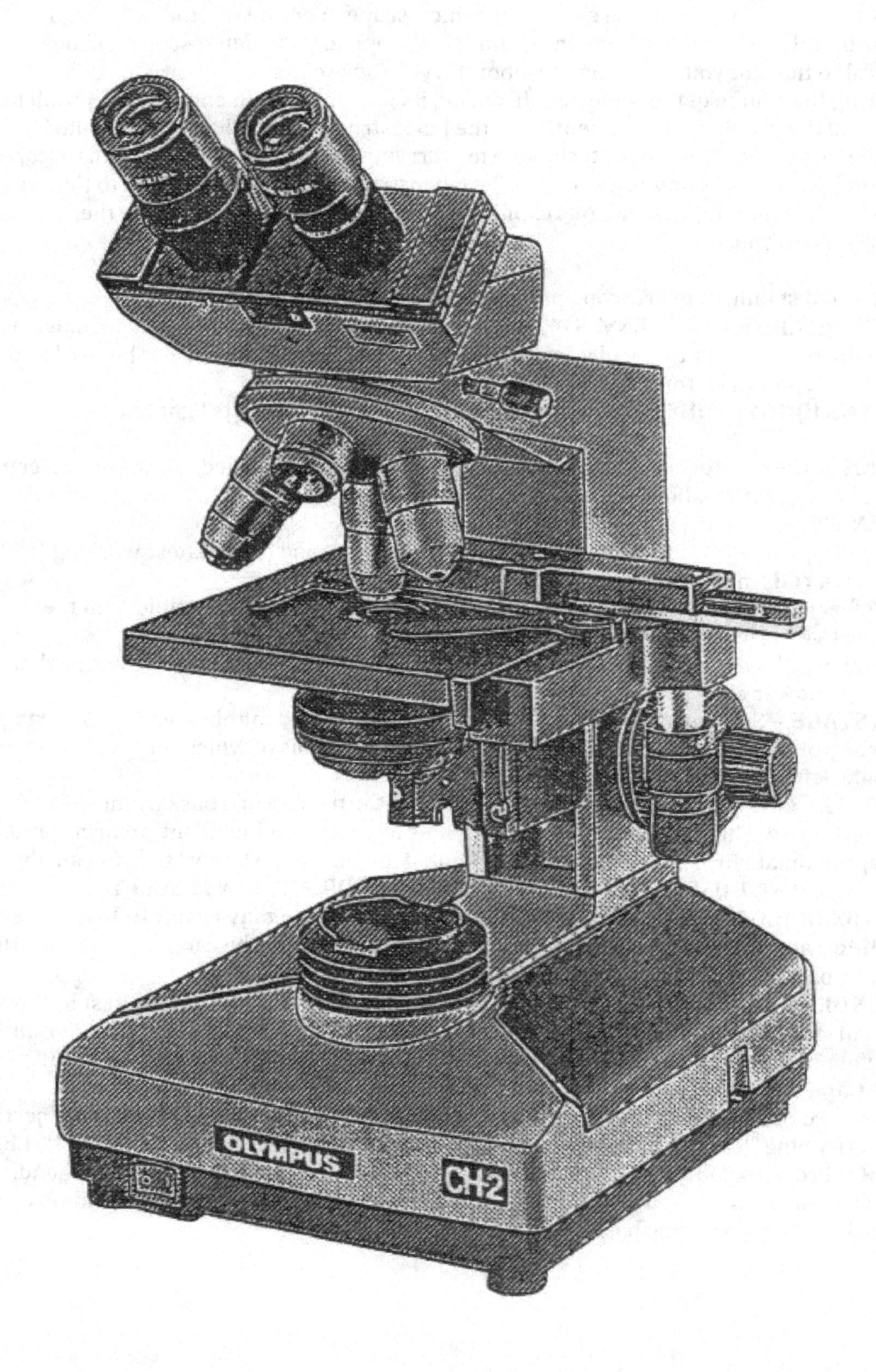

3. Obtain a prepared slide of the letter "e". Place the microslide on the mechanical stage with the slide label **to your left**.
4. While watching from the side of the microscope, use the coarse adjustment knob to slowly raise the mechanical stage until the end of the 4X objective is just above -- !but not touching! -- the microslide.
5. Look through the ocular lenses while adjusting the substage condenser and/or iris diaphragm to provide an appropriate amount of light.
6. While looking through the ocular lenses, slowly raise the mechanical stage using the coarse adjustment knob until the letter **comes into view**.
7. Looking only through the right-hand ocular lens, use the fine adjustment knob to sharpen the image of the letter **e**.
8. Looking through the left-hand ocular lens, use the diopter adjustment of the ocular lens to fine-focus the image.
9. Looking through the ocular lenses, adjust the substage condenser through an entire cycle, observing the effect upon the object under study -- in this case, the letter **e**.
10. Adjust the substage condenser to an optimum position. Normally, this is just slightly "below" its maximum upper position.
11. Looking through the ocular lenses, adjust the iris diaphragm through an entire cycle, observing the effect upon the object under study -- in this case, the letter **e**.
12. Adjust the iris diaphragm to an optimum position for observation.
13. Note the orientation of the letter **e** on the microslide without looking through the microscope (slide label to **your left!**). Draw the image below.

14. Now observe the orientation of the letter **e** as it appears through the microscope (slide label to **your left**!). Draw the image below.

Describe the different appearance of the letter **e** as viewed without and with the microscope.

15. Looking through the ocular lenses, use the mechanical stage to move the microslide **your right**.

In which direction does the image appear to move, left ( ) or right ( )? Inversion, a characteristic of compound light microscopes, refers to the fact that the image as viewed through the microscope is **inverted** and **reversed**. You must keep this feature in mind when adjusting the position of the microslide in respect to the position of the object under study.

Compound light microscopes are designed to be **parfocal,** once an object is in focus with the 4X objective, it should also be in focus with the 10X objective, and subsequent to refocusing with the 10X objective, it should be in focus with the 40X objective.

16. Looking through the ocular lenses, use the mechanical stage to move the microslide until the image of the letter e is **centered** in the field of view of the 4X objective and is in **sharp focus.**

Carefully rotate the revolving nosepiece until the 10X objective is in position directly above the opening in the mechanical stage. Refocus the image of the letter e.

On your previous drawing (page 3) of the letter e as observed with the 4X objective, draw a circle around that portion of the letter e as now seen through the 10X objective.

17. What do you conclude regarding the relationship between the size of the field of view and the power of objective lens in use?

18. With the image of the letter e in sharp focus and in the center of the field of view of the 10X objective, **carefully** rotate the revolving nosepiece until the 40X objective is in position directly above the opening in the mechanical stage.
19. **Using the fine adjustment knob ONLY**, bring the image of the portion of the letter e now in the field of view into sharp focus.

Do you see more ( ) or less ( ) of the image of the letter e with the 40X objective as compared to that observed when using the 4X or 10X objective?

NOTE: When beginning your study of a microslide specimen, **always** start with the 4X objective in position and move subsequently, as necessary, to the 10X and 40X objectives as outlined above in steps 16 - 19.

20. When you have finished your observation of the microslide of the letter e (**or any microslide for that matter**), carefully rotate the revolving nosepiece until the 4X objective is directly above the opening in the mechanical stage.
21. Using the coarse adjustment knob, lower the mechanical stage to its lower-most position. Remove the microslide from the stage and return it to its proper place as designated by your lab instructor. **Always follow this sequence in removing a microslide from the stage**

NOTE: During most lab sessions, you will be required to examine several different microslides. Take ONLY ONE microslide at a time to your work area; when finished, return the microslide before taking the next for study!

Total Magnification

Calculate the total magnification of the 4X, 10X and 40X objective lenses of your microscope in the space below.

Depth of Focus

1. **Obtain a microslide** with three different threads of different colors mounted one on top of another.
2. Using the 4X objective, locate a point where the three threads lie across one another. After securing focus with coarse adjustment, **slowly** focus "up and down" with the fine adjustment knob, noting that when a thread of one color is in sharp focus, the other two are somewhat blurred. You have observed the effect of depth of focus.
3. The vertical thickness of an object under observation, which is in focus at a given position of the microscope's body tube, represents the depth of focus for that particular lens or lens combination.
4. Now change to the 10X objective and observe the colored threads again.
5. What is the relationship of focus depth to the magnification power of the lenses in use?
6. Because of the characteristic of focal depth, you should routinely -- **but carefully** – focus "up and down" with the fine adjustment knob when studying an object with the microscope. Careful interpretation can thus give you an impression of the three-dimensional organization of the object under study.

Permanent Slide

1. It is often preserve and prepare a specimen for microscopic observation by making a permanent slide.
2. Obtain a slide of a single leaf of the aquatic vascular plant, **Elodea (Anacharis canadensis)**. Sketch a group of 3-4 leaf cells.
3. Observe the chloroplasts in the cells. Make sure you use 4X, 10X, and 40X to view the specium.
4. Observe the cells under **low** and **high** power. Make a drawing of a stained Elodea cell (at 40X) in your lab **on page 12**. Label the **central vacuole, nucleus, nucleolus, chloroplasts, cell wall,** and **cell membrane**. Do not spend an inordinate amount of time looking for the nucleus, as it can be difficult to discern.
5. Clean up the lab area so that it looks neat and tidy.

At the conclusion of each lab period, use the following procedure to prepare your microscope before leaving the lab:

1. Using a piece of lens paper, gently clean the ocular lenses and each objective lens.
2. Make sure the 4X objective is in place above the opening in the mechanical stage.
3. Set the light intensity control to "1"
4. Replace the electrical cord under the mechanical stage to prevent it from "hanging loose".
5. Replace the dust cover over the microscope.
6. Return the microscope to the proper shelf slot, making sure that the arm of the microscope is facing outward.

LABORATORY 2
STAGES OF MITOSIS

INTRODUCTION: Mitosis is an almost universal process by which plants and animals increase their number of cells. Cell division involves an equal division of the nucleus (**karyokinesis** - mitosis) and an equal division of the cell's cytoplasm (**cytokinesis**). During nuclear division, chromosomes are separated into two identical nuclei and mitosis is said to be an "**equatorial nuclear division**".

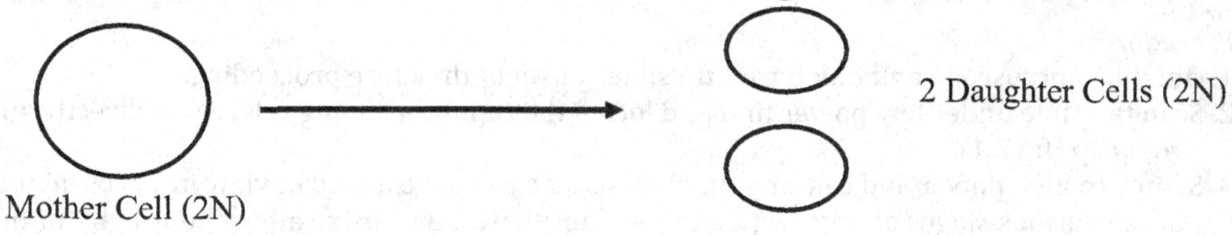

Mitosis and cytokinesis are parts of the life of a cell called the Cell Cycle. Most of the life of a cell is spent in a non-dividing phase called **Interphase**. Interphase includes **G1** stage in which the newly divided cells grow in size, **S** stage in which the number of chromosomes is doubled and appears as chromatin, and **G2** stage where the cell makes the enzymes & other cellular materials needed for mitosis.

Even though the entire process is a continuous one, karyokinesis is arbitrarily divided into five major steps for ease of study. In each of these stages, characteristic changes may be seen in the nuclear material. Before you begin your investigation, familiarize yourself with the names and characteristic appearance of plant cells in each phase. A summary is given below.

Interphase: The appearance of the cell is the same as a "typical," non-dividing cell. On a molecular level, however, chromosomes are replicating in preparation for nuclear division.

Prophase: Usually considered the first stage of active mitosis, prophase includes the disintegration of the nuclear membrane and the nucleolus. Also during this phase, the chromatin material shortens and thickens to become visible strands. Spindle fibers form, and by attaching to each replicated chromosome, help them migrate towards the center of the cell.

Metaphase: This phase is recognized as the time at which all the chromosomes have lined-up on the equitorial plane (mid-line) of the cell.

Anaphase: Each chromatid pair separates and is pulled to opposite ends of the cell by the spindle fibers. Each chromatid is now considered an entire chromosome. By the conclusion of this stage, two identical groups of chromosomes have migrated to either pole of the cell.

Telophase: During the last stage of mitosis, the chromosomes reform into interphase condition. Nuclear membranes and nucleoli reform around each group of chromosomes. The chromosomes themselves become long and indistinct. Spindle fibers disintegrate. Cell division usually occurs during telophase. In plant cells, a new cell wall, called a cell plate, forms along the equitorial plane. By the end of telophase, two identical daughter cells are formed. The cell plate between daughter cells is formed from the dictyosome (Golgi Apparatus).

Objective:

In this investigation, you will look at cells of a root tip, the rapidly growing area of a plant root. These cells will all be in various stages of mitosis. By counting the numbers of cells in each stage, it is possible to extrapolate the actual time needed to complete each phase. In general, the more cells you find in a specific stage of mitosis, the longer that stage takes to complete. Fewer cells in a given phase indicate a shorter period of time for completion.

Materials:
Microscope, prepared slide onion root tip, textbook, lab worksheet, pencil

Procedure:

1. Obtain a microscope and onion root tip slide. Clean both before proceeding.
2. Scan the slide under low power first and locate the rapidly growing cell region directly above the root cap (fig 2.1).
3. Switch to high power and center your slide so that you have a field of view in which all the cells are in various stages of mitosis (including interphase). Be sure to adjust your light for optimum viewing.
4. You are now going to identify the stage of each cell in your field. Starting at the top right corner of the field, record the stage of each cell in Data Table 1. Count your cells in a systematic manner. This will be considered area 1.
5. After completing the count in the first area, move your slide to a new area and perform the identification and count a second time. Record this data in the table as well.
6. Repeat the procedure a third time, with yet another field of view.
7. Make drawing of one cell in each of the various stages. Be sure to draw only what you actually see, but include all details that are visible.
8. Return to Data Table 2.1 and record the total numbers of cells in each phase, for all three trial areas.
9. Add the total number of cells viewed in each stage. Write the total count of cells viewed in all three trials in the appropriate place on the data table.
10. Using your data, calculate the time required for the completion of each stage. Be sure to use the totals for all three trials. Enter these results in the appropriate column of Data Table 1.
11. Prepare a **bar graph** to illustrate your results. The vertical **Y-axis** should be marked in "minutes to complete each stage." On the horizontal **X-axis**, allow equal space for each of the stages, beginning with interphase and ending with telophase.

There is a direct relationship between the number of cells counted in a given stage of mitosis and the time that that stage takes to complete. This may be calculated if the total time for mitosis in onion root tip cells is known. (The total time is measured from interphase to interphase). It is generally acknowledged that this time for onion cells is 720 minutes (12 hours). Set-up a ratio of the number of cells in each phase, compared to the total number of cells counted. Then multiply this fraction by the total time (720 minutes) needed to complete one mitotic division. Thus, the time for a specific phase is equal to:

Figure 2.1 Root apical meristem – region of cell division

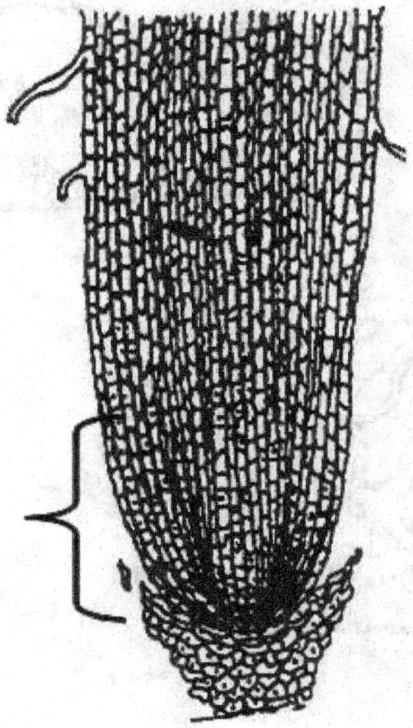

Figure 2.2 – Interphase,2. Interphase, 3. Early Prophase, 4. Prophase, 5. Late Prophase, 6. Metaphase, 7. Early Anaphase, 8. Late Anaphase, 9. Telophase, and 10. Cytokinesis

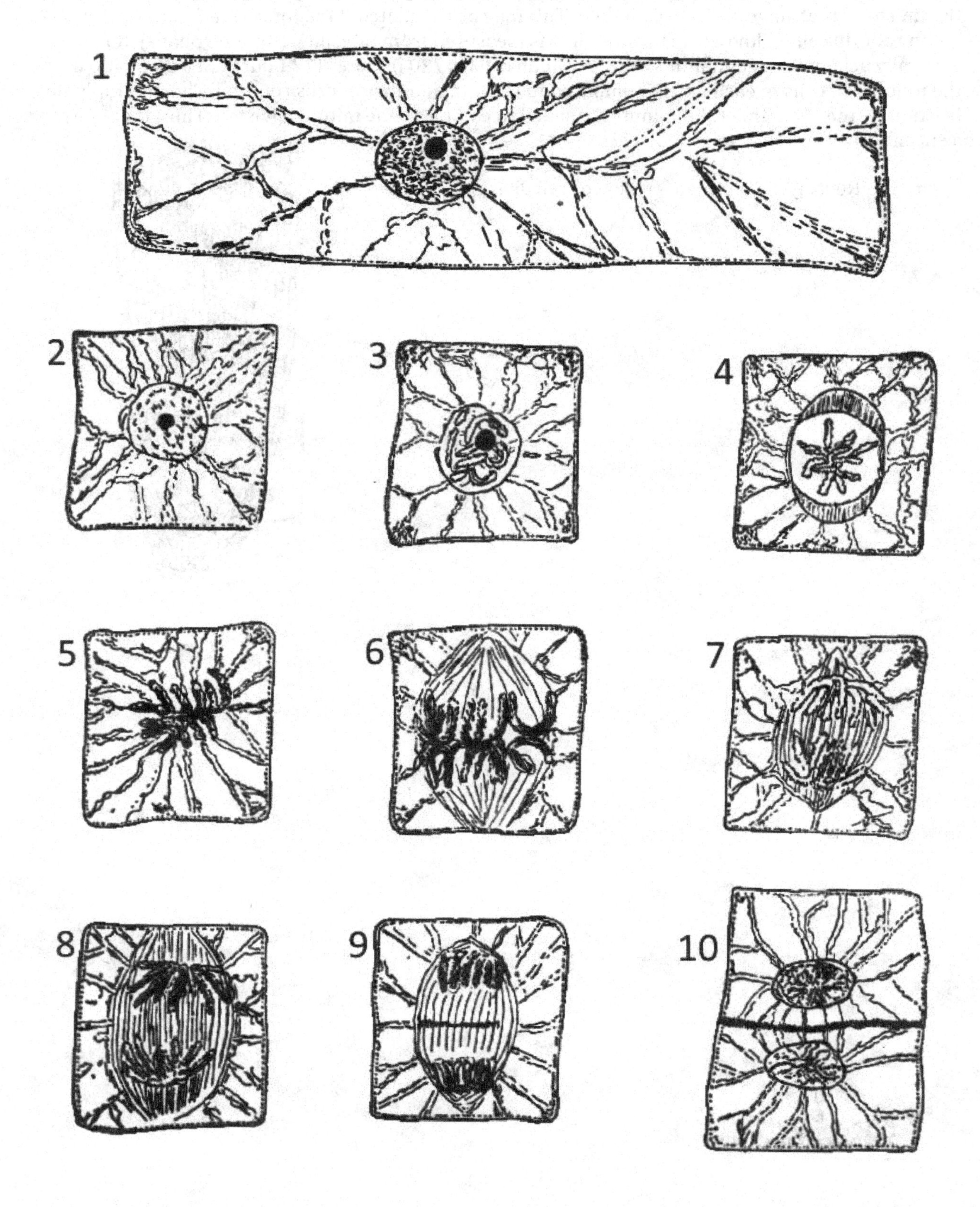

Label the successive stages of cell-division.

(a) Resting stage, showing: cytoplasm, nucleus, chromatin, nuclear membrane, nucleoli ; centrosome.

(b) Early division stage, showing: dissolution of nuclear membrane ; formation of spindle between two daughter centrosomes; spireme.

(c) Later division stage, showing: chromosomes arranged on spindle.

(d) Succeeding stage, showing: chromosomes divided and moving apart.

(e) Final stage, showing cell plate formation.

DATA TABLE 2.1

Stage of Mitosis	# cells in area 1	# cells in area 2	# cells in area 3	Total # Cells	$Time(\text{min}) = \frac{\# cells}{Total\,\# cells} * 720$ min
Prophase					
Metaphase					
Anaphase					
Telophase					
Interphase (Not a Mitotic Stage)					
Total # cells					

Questions:

1. Of the four stages of mitosis, which one takes the most time to complete?

2. Which is the shortest stage in duration?

3. What would happen if the process of mitosis skipped metaphase? telophase?

Title: ______________________________________

LABORATORY 3.
PHOTOSYNTHESIS

Photosynthesis, **the most important biochemical process occurring on Earth**, is a light-activated process in which the radiant energy is captured by a unique biochemical system. The absorbed light energy "excites" the photosensitive pigments of the chloroplasts with the energy of excitation being utilized to generate ATP and NADPH. Energy carried by the latter is subsequently utilized to reduce carbon dioxide to carbohydrate. Thus, radiant energy is converted into chemical energy in the bonds of the ATP and carbohydrate produced. For that reason, those organisms on Earth, which exhibit photosynthesis (or chemosynthesis) are often referred to as "primary producers" or "producer organisms". The process of photosynthesis, as exhibited by all photoautotrophic eukaryotes as well as the prokaryotic cyanobacteria and chloroxybacteria, may be summarized, in a greatly simplified manner, by the following equation:

$$6CO_2 + 12HOH \rightarrow [(CH_2O)]_6 + 6O_2 + 6HOH$$

NOTE: Work in-groups of two (2) on the following exercises.

THE LIGHT DEPENDENT REACTIONS OF PHOTOSYNTHESIS

You will recall from previous coursework that photosynthesis occurs in two linked stages. The initial series of reactions -- the **light dependent** reactions -- occur in the chloroplast grana and may be summarized:

$$HOH + ADP + Pi + NADP^+ \xrightarrow[chloroplast]{} \xrightarrow{light} O_2 \uparrow + ATP + NADPH + H^+$$

Radiant energy is utilized to energize the electrons of chlorophyll molecules and to split water (photolysis). Some of this energy is captured and used to form ATP and NADPH (reduced NADP); oxygen is released as a by-product. Because oxygen is released as a by-product, this type of photosynthesis is often referred to as **oxygenic photosynthesis**.

THE CARBON REDUCTION REACTIONS OF PHOTOSYNTHESIS

It was previously noted that photosynthesis occurs in two linked stages. In the first stage **light dependent** reactions -- energy from the Sun is captured in ATP and NADPH. During the second stage -- the **carbon reduction** reactions -- much of this energy is then stored in carbohydrates. The latter reactions, which occur in the chloroplast stroma, are sometimes inadvisably called the reactions **light independent** reactions of photosynthesis. The equation for the production of a molecule of the 6-carbon monosaccharide, glucose, may be summarized:

$$CO_2 + NADPH + H^+ + ATP \rightarrow C_6H_{12}O_6 + NADP^+ + ADP + P_i + HOH$$

The carbohydrate is normally "packaged" as the disaccharide, sucrose ($C_{12}H_{22}O_{11}$), for transport from the photosynthetic tissue to those tissues of the plant where the cells do not contain chlorophyll and, thus, cannot photosynthesize. For longer-term storage, the polysaccharide, starch, a polymer of glucose, is synthesized and stored inside the chloroplast and elsewhere in the plant. It has been indicated that the **light dependent** reactions must occur before **carbon reduction** reactions can occur. Can chlorophyllous cells/tissues carry out **carbon reduction** if light is "blocked", interfering with the **light dependent** reactions? Is light, indeed, necessary for photosynthesis?

Release of Oxygen During Photosynthesis and the Effect of Light Intensity Upon Photosynthetic Rate

Photosynthesis—If the leaf of a plant which has been exposed to sunlight is killed, decolorized, and treated with a solution of iodine, it becomes blue or blue-black in color. This reaction occurs only when starch or starch-like materials are treated with iodine, and we may conclude that the leaf contains starch. The plant has no outside source of starch; no test for starch can be secured in the soil in which it grows or in the air which surrounds it; the starch is made within the plant. In fact, the starch is made from sugar and the sugar from the gas, carbon dioxide, and from water. The sugar, glucose, is probably the first synthetic product of this reaction which can be demonstrated in the plant, but it is rapidly changed, in most plants, to starch. We, therefore, justifiably use the simple iodine test for starch to demonstrate the occurrence of photosynthesis in most plants.

Experiment

By suitable experiments it may also be shown that oxygen appears within the leaf at the same time that sugar or starch does, but is given off into the air. A simple way of showing this is to expose a water plant like Elodea to strong light. The oxygen as it is given off is easily visible, for in the water it forms small bubbles. If we wish we may collect these bubbles by placing the Elodea under an inverted funnel, using sufficient water to cover the funnel. If we now invert a test tube filled with water over the end of the funnel, the gas bubbles as they rise will displace the water and collect in the tube. If we place a thumb over the end of the test tube before we lift it from the water and then, as we remove our thumb, insert into the tube a glowing splinter, it will glow more brightly or burst into flame. This indicates the presence in the tube of more oxygen than there is in the air, since oxygen is the only gas which supports combustion.

Figure 3.1 Diagram of oxygen collection apparatus.

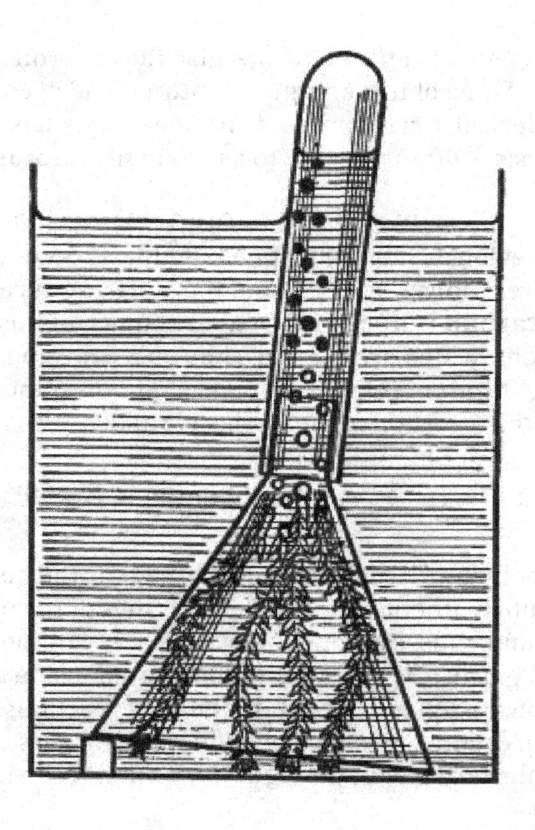

Light and Its Absorption by Chloroplast Pigments

"White" light is composed of all the wavelengths in the visible portion (**400 - 700 nm**) of the electromagnetic radiation spectrum. Typically, these wavelengths are measured in nanometers (nm). We perceive colors because materials contain photoreceptive pigments that selectively and characteristically absorb some wavelengths of visible light while reflecting and/or transmitting other wavelengths. Thus, the color of a material is due to the wavelengths that are reflected and/or transmitted by that material, the wavelengths absorbed by that material!

The various photoreceptive pigments found in the chloroplasts of plants -- including the primary photoreceptor, chlorophyll **a** (C_a), and the accessory pigments, Chlorophyll b (C_b), carotenes and xanthophylls -- absorb different wavelengths of radiant energy, making use of a wide range of light energy that can be captured for photosynthesis.

Paper Chromatography of Leaf Pigment Extract

1. Obtain a strip of chromatography paper, which has been cut to size.
2. Using the applicator supplied, apply a small drop of leaf pigment extract supplied by the lab instructor in the center of the paper strip about 2.5 cm (about 1 inch) from one (end the bottom). Allow the spot to dry for 45 seconds. Repeat this process four times, taking care to apply each succeeding drop of pigment extract over the proceeding location.
3. Using the Erlenmeyer flask into which has previously been placed chromatography solvent, carefully place the paper strip into the flask. Be sure that the bottom of the paper strip extends into the solvent **without the area with the pigment extract doing so.**
4. Replace the piece of plastic wrap over the top of the flask to prevent solvent fumes from permeating the lab. **Do not agitate the flask contents**!
5. When the solvent front reaches a point about 10 cm (about 4 inches) from the bottom of the paper strip, remove the strip from the flask and allow it to dry for 1 - 2 minutes. **Replace the plastic wrap over the flask.** NOTE: The solvent contains substances that are both toxic and flammable. Handle with care!
6. Observe the pattern of pigment separation on the paper strip. You should see a faint yellow orange band of carotenes nearest the solvent front. At sequentially lower levels, you see a pale-yellow band of xanthophylls, a bluish-green band of Chl **a** nearest the bottom, an olive-green band of Chl **b**. In the space below, sketch the appearance of the chromatography strip labeling the pigment bands.

Demonstration

Observe the test tube containing some of the leaf pigment extract by holding it up to the ceiling lights. What color is the solution by transmitted light? ________________. Why do you not see the colors of the yellow-orange carotenes and pale-yellow xanthophylls whose presence was revealed by paper chromatography?

Absorption Spectrum of Leaf Pigment Extract

Having demonstrated the presence of Chl **a**, Chl **b**, carotenes and xanthophylls extracted from plant leaves, you are to determine the wavelengths of light which are maximally absorbed (minimally transmitted) by the predominant pigment, chlorophyll, present in the pigment extract. For this purpose you will employ a spectrophotometer (Spectronic 20), an instrument you have used in previous courses. **If you need assistance in the proper use of the Spectronic 20, ask your lab instructor.**

Obviously, chlorophyll predominates in the leaf pigment extract. Though the other pigments present will influence the absorbance/transmittance data you acquire with the Spectronic 20, the data will largely reveal the absorbance/transmittance characteristics of the predominant pigment, chlorophyll. Each group has been provided with three spectrophotometer tubes containing: (1) a strip of white paper; (2) acetone [the solvent]; and (3) a dilute solution of leaf pigment extract. NOTE: Keep the foil cover on the acetone blank and leaf extract tubes al all times **except** when placed into the sample holder of the Spectronic 20.

1. Set the wavelength selector knob on the Spectronic 20 at 660 nm.
2. Insert the spectrophotometer tube containing a strip of white paper into the sample holder; leave the sample holder cover open. Looking down into the tube, what color do you see? _______________.
3. Change the wavelength selector knob to 600 nm. What color do you see? _________.
4. Change the wavelength selector knob to 560 nm. What color do you see? _________.
5. Change the wavelength selector knob to 440 nm. What color do you see? _________.
6. Change the wavelength selector knob to 400 nm. What color do you see? _________.
   You will have noted that, as you vary the wavelength of light produced by the Spectronic 20, you vary the wavelength of light available for absorption by a material contained within the sample holder.

7. The chloroplast pigments of the leaf extract are in an acetone solution. Since the solvent itself absorbs some light, this amount must be accounted for by "subtracting" it from the total absorption for the pigments alone. This is done by using the tube containing the acetone only (solvent blank). Set the wavelength selector knob to 380 nm. With spectrophotometer tube in the sample holder but with the sample holder cover **closed**, use the left-hand, front knob on the Spectronic 20 to adjust to inifinite absorbance (0% transmittance).
8. Place the acetone blank into the sample holder, close the sample holder cover and use the right-hand, front knob on the Spectronic 20 to adjust to zero absorbance (100% transmittance). Remove the acetone blank from the sample holder.
9. Place the spectrophotometer tube containing the dilute leaf pigment extract into the sample holder, close the sample holder and record -- in the Table below – the absorbance and transmittance values at 380 nm.
10. Determine and record the absorbance and transmittance values, successively, @ 400, 4420, 440, 480, 560, 600, 620, 640, 660, 680, 700 and 720 nm. Note: rezero the instrument using the solvent blank time you change the wavelength setting (above).
11. With the foil covers in place on the acetone blank and leaf pigment extract tubes, return the spectrophotometer tubes to you lab instructor.

Table 3.1. Absorbance and Transmittance Values of Leaf Pigment Extract

Wavelength (nm)	Absorbance	Transmittance (%)
400		
420		
440		
480		
560		
600		
620		
640		
660		
680		
700		

Using a sheet of graph paper supplied by the lab instructor, plot the data with wavelength on the horizontal axis, absorbance on the left-hand vertical axis and transmittance on the right-hand vertical axis.

a. **At what wavelength(s) did you detect maximum absorption? The wavelength(s) correspond(s) to what color(s) in the spectrum?**

b. **At what wavelength between 400 and 700 nm did you detect minimum absorption? That wavelength corresponds to what color in the spectrum?**

c. **At what wavelength(s) and corresponding color(s) is chlorophyll excited (activated) for photosynthesis?**

d. **Why is chlorophyll green in color?**

e. **During the spring and summer in our area, the leaves of most plants are obviously green. Why do the leaves exhibit other colors in the fall of the year?**

Note: Chl **a** has absorption maxima @ 430 and 663 nm; the "peak" @ 430nm
Chl **b** has absorption maxima @ 435 and 645 nm; the "peak" @ 435nm

THE CARBON REDUCTION REACTIONS OF PHOTOSYNTHESIS

It was previously noted that photosynthesis occurs in two linked stages. In the first stage **light dependent** reactions -- energy from the Sun is captured in ATP and NADPH. During the second stage -- the **carbon reduction** reactions -- much of this energy is then stored in carbohydrates. The latter reactions, which occur in the chloroplast stroma, are sometimes inadvisably called the reactions **light independent** reactions of photosynthesis. The equation for the production of a molecule of the 6-carbon monosaccharide, glucose, may be summarized:

$$CO_2 + NADPH + H^+ + ATP \rightarrow C_6H_{12}O_6 + NADP^+ + ADP + P_i + HOH$$

The carbohydrate is normally "packaged" as the disaccharide, sucrose ($C_{12}H_{22}O_{11}$), for transport from the photosynthetic tissue to those tissues of the plant where the cells do not contain chlorophyll and, thus, cannot photosynthesize. For longer-term storage, the polysaccharide, starch, a polymer of glucose, is synthesized and stored inside the chloroplast and elsewhere in the plant. It has been indicated that the **light dependent** reactions must occur before **carbon reduction** reactions can occur. Can chlorophyllous cells/tissues carry out **carbon reduction** if light is "blocked", interferring with the **light dependent** reactions? Is light, indeed, necessary for photosynthesis? Four treatments, described below, of the leaves of geranium plants (**Pelargonium**) will be used to address these questions.

NOTE: For the following, --A thru D--, work in groups as designated by your lab instructor. Share your results with other groups in the lab.

A.

1. Obtain a leaf from a geranium plant that has been kept in the light and has **shield** on the leaf.
2. Immerse the leaf in a beaker of hot water from about 1 minute to disrupt the cell and chloroplast membranes.
3. Using forceps, transfer the leaf to a beaker of hot 95% ethyl alcohol. **NOTE: Alcohol boils at a lower temperature than does water and is flammable. Do not heat alcohol directly with an open flame and keep any flames away from alcohol fumes!**
4. Leave the leaf in the alcohol until virtually all the chlorophyll has been removed, about 4 - 5 minutes – the leaf appears "whitish" throughout.
5. Using forceps, transfer the leaf to a dish of cool water for about 15 seconds.
6. Using forceps, transfer the leaf to a culture dish; "unfold the leaf so that it lies flat in the dish", (using dissecting needles as necessary), **Leave the leaf in the IKI for 3 minutes.**
7. Using forceps, transfer the leaf to a culture dish specifically labeled for the treatment.
8. In the space below, sketch the leaf in outline, indicating by shading those parts of the leaf in which the starch, if any, is localized. **The presence of starch is indicated by a dark color!**
9. Leave the leaf in the dish for examination by other groups.

B.

1. Obtain a leaf from a geranium plant that has been kept in the light for four days with a portion of the leaf having been covered with a "dark shield".
2. Treat the leaf in the manner outlined above in "-- steps 2 - 9".

C.

1. Obtain a leaf from a geranium plant that has been kept in the dark for four days and has **no** "dark shield" on the leaf.
2. Treat the leaf in the manner outlined above in "-- steps 2 - 9".

D.

1. Obtain a leaf from a geranium plant that has been kept in the dark for four days with a portion of the leaf having been covered with a "dark shield".
2. Treat the leaf in the manner outlined above in "-- steps 2 - 9".

Compare the sketches you have made of the four leaves. What can you conclude regarding the relationship between light and the **light dependent, carbon reduction** stages of photosynthesis?

LABORATORY 4.
ALTERNATION OF GENERATIONS

Introduction

Plants are separated into different phyla based largely on differences in their reproductive lifecycles. To understand these differences, you must first understand the basic underlying reproductive lifecycle common to all plants, known as **ALTERNATION OF GENERATIONS**. Alternation of generations is the most difficult material of the plant portion of this web site. Once you understand alternation of generations, the rest will be easy!

All plants and most algae have a sexual life cycle that consists of an alternation between a haploid (n= one copy of each chromosome) gametophyte generation, and a diploid (2n= two copies of each chromosome) sporophyte generation. The haploid gametophyte generation produces haploid gametes (eggs and sperm), which unite to form a diploid zygote. The gametophyte is usually multicellular, while the gametes and the zygote are single cells. The zygote grows by mitosis into the diploid multicellular sporophyte. The diploid sporophyte contains specialized regions where meiosis occurs, creating haploid spores. The spores, once dispersed, germinate into haploid gametophytes, starting the cycle anew. Some of the differences that will be found between plant phyla include relative dominance, size, and independence of each of the two generations. Make sure you understand the basic concept of alternation of generations before proceeding, as you will be learning concepts that build on your understanding of this basic sexual reproduction cycle.

STUDY HINT:

- the **gameto**phyte generation always produces **gametes**
- the **sporo**phyte generation always produces **spores**

Forms of Reproductive Life Cycles

The two extremes:

- In the **haplontic** (or haploid dominant) life cycle, after the gametes (n) fuse to form the zygote (2n), the zygote immediately undergoes meiosis to produce spores (n), which regenerate the gametophyte generation (n). Thus, only one single cell, the zygote, is ever diploid. This type of life cycle is seen in most some green algae.
- In the **diplontic** (or diploid dominant) life cycle, on the other hand, ALL cells are diploid, with the exception of the gametes. This life cycle will be familiar to you since it is what is seen in humans and most other animals: only the egg and sperm are haploid, all other cells are diploid. The diplontic life cycle is not seen in most algae but is seen in most terrestrial plants.

The various life cycles seen in most of the algae and in plants are somewhere in between these two extremes. In these cases both the sporophyte (2n) and gametophyte(n) generations are multicellular. Sometimes they are indistinguishable from each other in any way except for their chromosome number (eg. some species of Phaeophytes), while in other cases either the sporophyte or the gametophyte generation will be larger and more familiar than the other.

Alternation of Generation in Nontracheophytes – nonvascular plants (Liverworts, Hornworts, and Mosses)

In non-tracheophytes, the gametophyte generation is the dominant generation, and is the more prominent green "leafy" structure. This **gametophyte** (n) produces gametes (n) in multicellular **gametangia**. If the gametangia is female, and produces eggs (n), it is known as the **archegonium**, while if it is male and produces sperm (n), it is known as the **antheridium**. Sperm are released from the antheridium and swim to the archegonium, where they fertilize the egg to form the zygote (2n). The zygote develops into the sporophyte embryo (2n), which as it grows eventually becomes nutritionally independent from the gametophyte. Mature **sporophytes** (2n) produce **spores** (n) by meiosis in **sporangia**; the spores are eventually dispersed by the wind. When a spore germinates, it grows into another gametophyte (n), starting the cycle all over again.

Alternation of Generation in Non-seedbearing Tracheophytes
(Club Mosses, Horsetails, Ferns)

All nonseed tracheophytes reproduce by means of spores. The sexual reproduction lifecycle shows an alternation of generations between gametophyte and sporophyte generations just as in the nontracheophytes, however, in the tracheophytes (nonseed and seed), the diploid sporophyte generation is dominant, not the gametophyte generation. The leafy green plant you think of when you think of ferns, for example, is the sporophyte generation. Gametophytes of nonseed tracheophytes are very tiny, but are free living (live separately from the parent sporophyte). Water is still required for the sperm to reach the egg in nonseed tracheophytes.

Reproduction in Seed-bearing Tracheophytes
(Gymnosperms and **Angiosperms)**

Today, seed bearing plants vastly outnumber the nonseed plants. These species are very successful for several reasons.

- First, they have bypassed the need to have water in their environment for fertilization to occur, by the formation of **pollen**; very small male gametophytes that are transported to the female gametophytes by wind or insects. Thus the sperm, instead of having to swim to the egg, is brought to the egg by the male gametophyte. Upon arrival the male gametophyte produces sperm, which are carried to the egg inside its protective structure (the **archegonium**), by a pollen tube.

- Second, the spore (which was the means of dispersal in nonseed tracheophytes) is replaced by the **seed** as the means of dispersal. Seeds consist of a seed coat, the embryo, and stored food. Seeds protect the young sporophytes and provide them with food reserves.

- Finally, the evolutionary trend has been towards reduction of the gametophyte to being of extremely small size, and dependent on the sporophyte for nutrition.

Reproduction in Gymnosperms

Gymnosperms include cycads, gingkos, conifers, and gnetophytes. All of these species produce seeds that are covered by a seed coat but not a fruit. These species DO NOT produce flowers. All produce pollen, which can be transported either by the wind, or by wind and insects in some species. Pollen makes it unnecessary for the plant to live in a moist environment; the sperm no longer needs water to swim to the egg, rather the whole male gametophyte (pollen grain) is transported to the female. Gymnosperm gametophytes are smaller than those in the nonseed tracheophytes, and are not free-living. This reproductive cycle still shows alternation of generations, however the haploid gametophyte is very reduced in size and no longer lives independently of the sporophyte.

Reproduction in Angiosperms

In flowering plants the gametophyte generation is more reduced than in any other plant phylum. The male and female gametophytes are composed of even fewer cells than in gymnosperms, and are completely dependent on the sporophtye generation. The male gametophyte is the germinated pollen grain. Unlike in other plant phyla (except **Gnetophyta**), TWO male gametes, both contained within a single pollen grain, participate in the fertilization process. One male gamete (sperm) combines with the egg to form the zygote, while the other combines with two haploid nuclei of the female gametophyte, to form a triploid nucleus. This triploid nucleus divides to form **endosperm**. Endosperm is a nutritive tissue formed only in seeds of angiosperms. It stores starch, lipid reserves, and other compounds for use by the developing sporophyte. Fertilization that involves two male gametes is known as **double fertilization**.

After fertilization in angiosperms, the **ovary**, together with its seeds, develops into a **fruit**. Fruits are formed only in angiosperms. A fruit can consist only of the mature ovary and its seeds, or it can include other flower parts. Many fruits play a role in seed dispersal. Animals eat many fruits, the seeds of which travel unharmed through their digestive systems, and usually deposited away from the parent plant, along with a good supply of fertilizer! Other fruits (eg. dandelions, milkweed) are carried on the wind to a new location, aided by specialized parts of the fruit that enable it to be carried by the wind. Other fruits have specialized hooks and barbs which allow them to stick to animal fur, to be carried with the animal until they finally fall off or are taken off by the animal (eg. burdock, tick-seed clover).

LABORATORY 5.

THE ALGAL GROUPS -- OXYGENIC, PHOTOAUTOTROPHIC PROKARYOTES & PROTISTS

Though the term, alga (pl., algae), is no longer used in a formal nomenclatural sense, it is still useful to identify several distantly, if at all, related groups which share certain features:

1) They are all photoautotrophic; all possess Chl **a** as the primary photoreceptive pigment and evolve oxygen as a by-product of photosynthesis, (i.e., **they are oxygenic photoautotrophs**).
2) Though some may exhibit rather complex body organization, none have achieved tissue/organ level organization with the presence of specialized vascular tissues, xylem and phloem.
3) Their reproductive structures -- gametangia and sporangia -- are unicellular or, if multicellular, the structures do not possess sterile jacket cells, (i.e., **all cells of a gametangium or sporangium are fertile**).
4) In reproducing sexually, none retain the zygote within the female gametangium for development into a multicellular embryo, (i.e., **they are non-embryophytic**).

Using the above defining features, many **phycologists** (biologists who specialize in the study of algae) place these organisms into one of eleven Divisions. Of these, two, the DIVISIONS CYANOBACTERIA and PROCHLOROPHYTA (=CHLOROXYBACTERIA), are placed into the KINGDOM PROKARYOTA (=MONERA) as they are prokaryotic. The remaining groups are all-eukaryotic and are placed into nine divisions of the KINGDOM PROTISTA: EUGLENOPHYTA, CRYPTOPHYTA, RHODOPHYTA, DINOPHYTA, HAPTOPHYTA, BACILLARIOPHYTA, CHRYSOPHYTA, PHAEOPHYTA and CHLOROPHYTA. Several features are utilized to segregate the eukaryotic algal groups from one another including, but not limited to: (1) the nature of their accessory photoreceptive pigments; (2) ultrastructural features of the chloroplast; (3) the chemical nature of their reserve food material; (4) the chemical nature of their cell walls, if present; (5) ultrastructural features and the number and insertion of flagella on motile cells, if any.

Regardless of one's preference for one algal classification system over another, they are important organisms as they are the predominant primary producers of most aquatic ecosystems, both marine and freshwater. Time and availability of material makes it impractical to study representatives of all the algal groups in the laboratory; only a selection of algae will be studied in the lab.

KINGDOM PROKARYOTA (MONERA) -- The Prokaryotic Algal Groups

DIVISION CYANOBACTERIA -- cyanobacteria; blue-green algae

In addition to Chl a and a variety of carotenes and xanthophylls, the cyanobacteria possess a series of water soluble photoreceptive pigments -- phycobiloproteins -- including c-phycocyanin and c-phycoerythrin. Unlike eukaryotic photosynthetic organisms, the photoreceptive pigments of the cyanobacteria are not localized within chloroplasts; rather, they appear as granules associated with free membranes (**thylakoids**) in the cytoplasm. Flagellated cells are totally lacking among the cyanobacteria, and none exhibit true sexual reproduction. The major means of reproduction is by an asexual process, **prokaryotic binary fission**. Body organization among these forms ranges from unicellular to colonial to filamentous. Among the filamentous forms, some are branched, others are unbranched.

Gloeocapsa is a common unicellular cyanobacterium whose cells, after binary fission, often remain adherent within a common mucilage envelope; thus, one often finds the organism occurring in groups of 2-4-8 cells. Study the prepared slide of this cyanobacterium. Locate and observe a cell or group of cells. Notice that the organization of the cell contents is such that you can discern an area more or less in the center of the cell that is more granular in appearance than is the peripheral region. The near-center granular region is the **nucleoid** ("centroplasm"), the site of localization of most of the cell's DNA. The peripheral region, the **pigmented cytoplasm**, is the area where the thylakoids tend to be concentrated. After initial observation, place a small drop of IKI to the edge of the coverslip; this will facilitate observation of the **cell wall, mucilage sheath**.

Locate some cells in various stages of binary fission and cohesion after cell division. Note the centripetal direction of cleavage and the **cleavage furrow**. Label the following figure.

Figure 5.1. *Gloeocapsa* species (A-E)

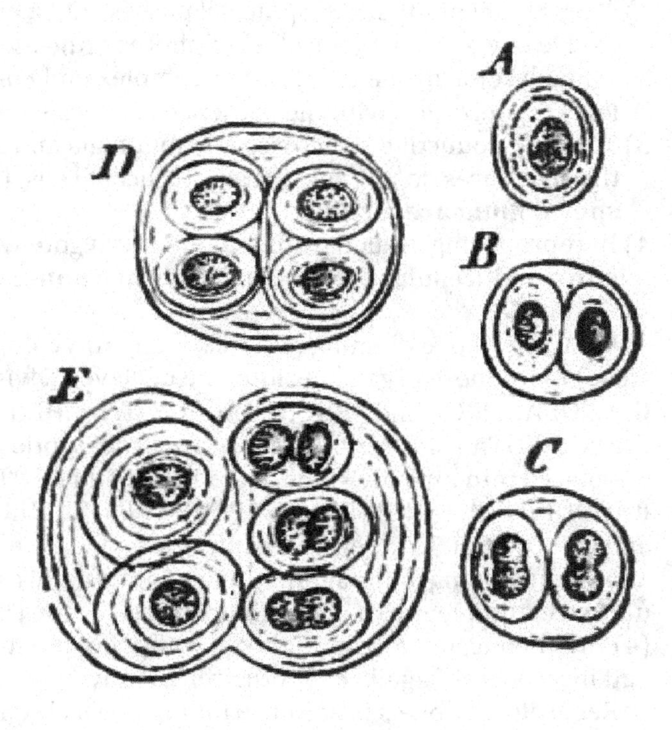

Anabaena is a common unbranched, filamentous cyanobacterium. In addition to ordinary vegetative cells, this and some other cyanobacterial taxa produce two types of modified cells, **heterocysts** and **akinetes**. Many cyanobacteria exhibit a very important biological function, **nitrogen fixation**, which is often associated with heterocysts. The akinete functions as an asexual reproductive spore and, much like the endospore of true bacteria, is highly resistant to adverse environmental conditions, especially desiccation.

Study the prepared slide of this cyanobacterium. You will note the common occurrence of heterocysts; do not spend an inordinate amount of time looking for an akinete as they are much less common. Label the following figure: **cell wall, vegetative cell, heterocyst, akinete**.

Figure 5.2. *Anabaena* species with heterocyst.

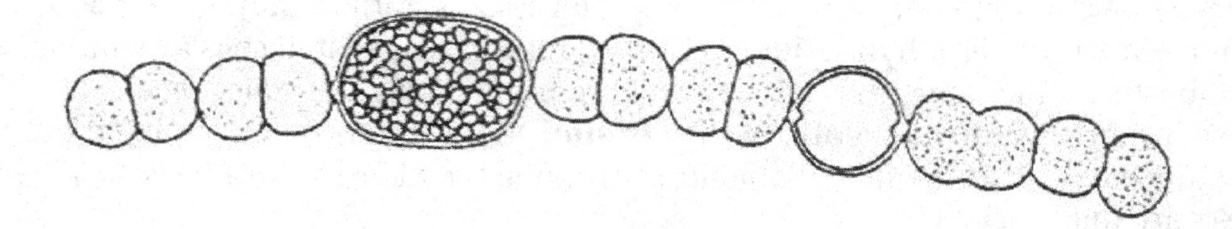

KINGDOM PROTISTA -- The Eukaryotic Algal Groups

DIVISION EUGLENOPHYTA – euglenoids

This algal group, most of which are flagellated unicells, are segregated from the other algal groups on, among other features, their lack of cell walls and production of **paramylum** as a reserve food material. Most have only one emergent flagellum. Though a cell wall is lacking, the cytoplasmic membrane is firm enough to give the cell a characteristic shape; a modified cytoplasmic membrane of this nature is often referred to as a **pellicle** or **periplast**. The pellicle of euglenoids may be variously ornamented with striations, granules, verrucae, etc.

Figure 5.3. *Euglena* sp. with vegetative cell, reproduction, and cyst.

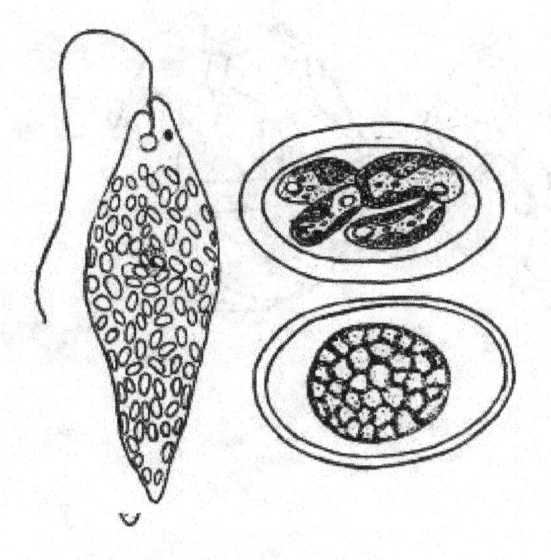

Species of ***Euglena*** are among the most common euglenoids.

Get a slide and label on the following figure: **pellicle, pellicular striation, stigma (red eye spot), flagellum, chloroplast**, and **paramylum body**. The pellicular striations and flagellum may be difficult to discern. The nucleus, which may be difficult to discern, is usually located "midway" between the two-paramylum bodies of this species Euglena. The nucleus appears as a generally spherical, "structure".

DIVISION DINOPHYTA – dinoflagellates

Most dinoflagellates are biflagellated unicells. Among other distinguishing features, the pattern of flagellar insertion is unique to this group with one flagellum, the **girdle flagellum**, lying in the **transverse groove**, and the second flagellum, the **trailing flagellum**, lying partially in the **sulcus** before trailing posteriorly.

Study the prepared slide and, if available, **make a wet mount** of this common dinoflagellate. Note the **transverse groove** and the patterned cellulosic cell wall plates; the latter identify this as one of the "armored" dinoflagellates. Label on the following figure: **cell wall, transverse groove**, and, if visible, the **sulcus**.

Figure 5.5. *Peridinium* with sulcus and transverse groove

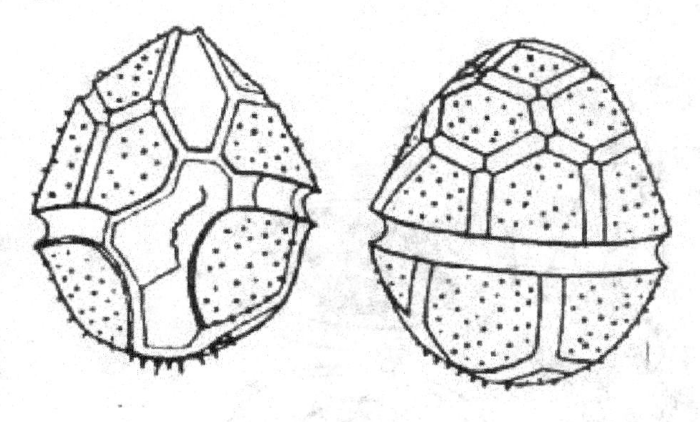

DIVISION BACILLARIOPHYTA – diatoms

Most diatoms are unicellular and some, though non-flagellated, move over aquatic substrates by a "gliding/sliding" motion, the exact mechanism of which is still poorly understood. Their cell walls are composed largely of silica and occur in two equal halves. One of the two cell wall halves, the **hypotheca**, fits inside the other half, the **epitheca**, as the two halves of a Petri dish. Diatom cell walls are distinctively ornamentated with striations, punctae, granules,

Demonstrations
Observe the prepared diatom type slide, which will give you an appreciation for the variety of cell shapes and ornamentation typical of these organisms. **Careful, please; this is a very expensive slide! Use fine adjustment only!**

Observe the prepared slide showing **girdle** and **valve views**, as well as a **dividing**, of a common freshwater diatom, ***Pinnularia***.

Figure 5.6. *Pinnularia* (A) valve view (B) girdle view

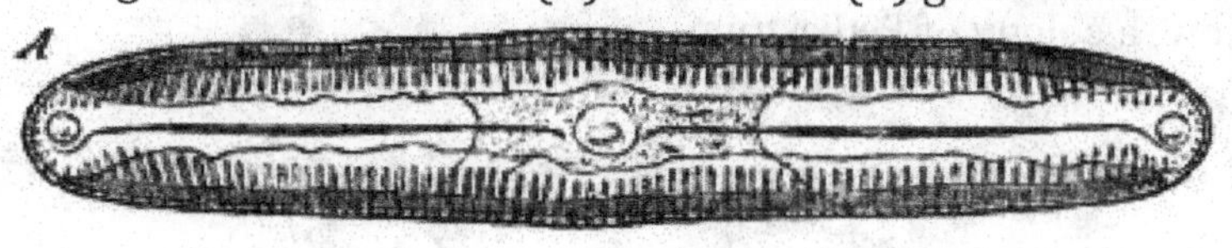

DIVISION CHLOROPHYTA -- green algae

The chlorophytes, a very large and diverse group, exhibit six types of body organization: (1) flagellated unicellular and colonial forms; (2) non-flagellated unicellular and colonial forms; (3) filamentous -- unbranched or branched -- forms; (4) membranous forms; (5) parenchymatous forms and (6) coenocytic and tubular [siphonous] forms. Time and availability of material does not permit study of all representative forms.

The DIVISION is comprised of two Classes, the Class **CHLOROPHYCEAE** sensu stricto (chlorophycean green algae) and the Class **CHAROPHYCEAE** sensu lato (charophycean green algae). It is the latter group that most botanists believe served as the progenitors of embryophytic plants.

CLASS **CHLOROPHYCEAE** sensu stricto

Chlamydomonas is a motile single-celled chlorophyte the cells of which bear two flagella. With careful observation, you should be able to see the two flagella and the stigma. **Study the prepared slide** of ***Chlamydomonas***.

Figure 5.6. Chlamydomonas species with two flagella.

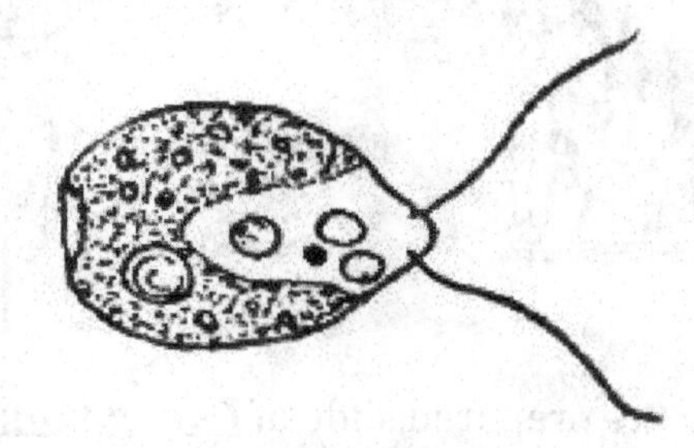

Pediastrum is a non-motile colonial chlorophyte. **Make a wet mount** from the culture provided. Study the **prepared slide** and sketch, in the space below, a colony of Pediastrum.

Figure 5.8. *Pediastrum* species

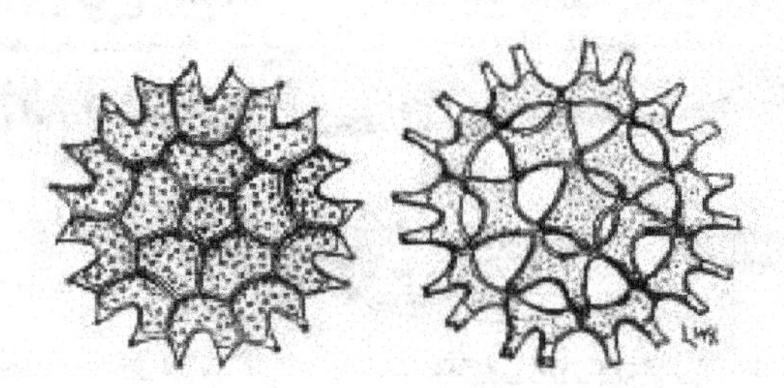

Oedogonium is a common unbranched filamentous chlorophyte. Species are oogamous; i.e., they produce large, non-motile female gametes within a female gametangium, the oogonium, and smaller, motile male gametes within a male gametangium, the antheridium. Observe the structure of a vegetative cell. Label Figure A. below: **cell wall**, **pyrenoid**, and, if visible, **nucleus**. After initial observation, the addition of a small amount of IKI to the edge of the coverslip may facilitate resolution of the nucleus. You will note that the **chloroplast** is a reticulate (net-like) structure.

Figure 5.9. (A) ***Oedogonium*** vegetative cells (B) reproductive structures

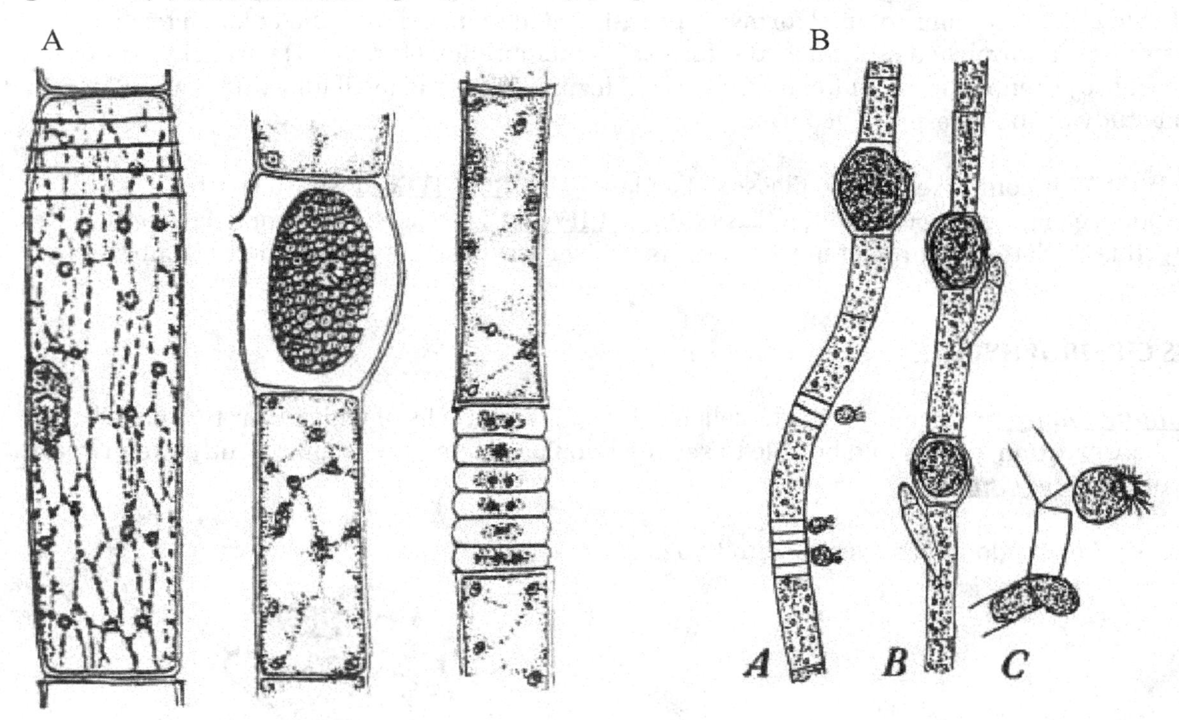

Study the prepared slide of ***Oedogonium*** to see the reproductive organs. Label Figure B. above: oogonium, antheridium

CLASS **CHAROPHYCEAE** sensu lato

Spirogyra is a very common unbranched filamentous charophycean green alga. Study the prepared slide Spirogyra Conjugation. The dominant life cycle stage is **haploid (1N).** Spirogyra will undergo sexual reproduction called Conjugation forming a **diploid zygote (2N).**

Note the spiral, band-shaped chloroplast(s). Label the following figure: **cell wall, chloroplast, pyrenoid, nucleus, zygote, and conjugation tube.**

Figure 5.10. (A) Spirogyra undergoing conjugation (B) Nucleus migration and zygote.

A

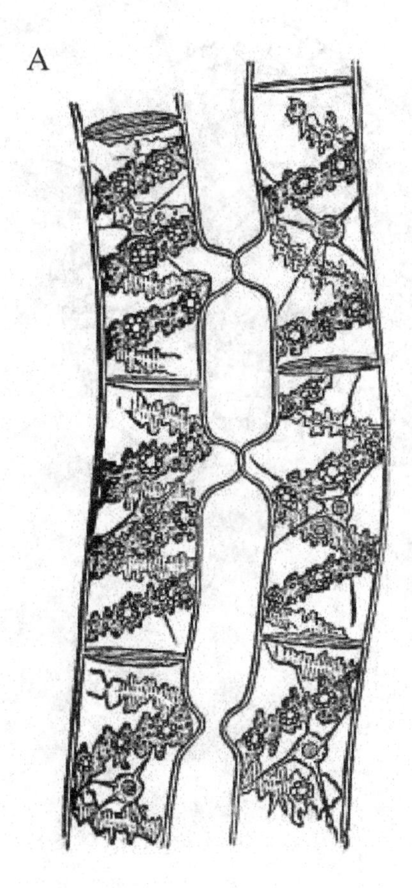

B

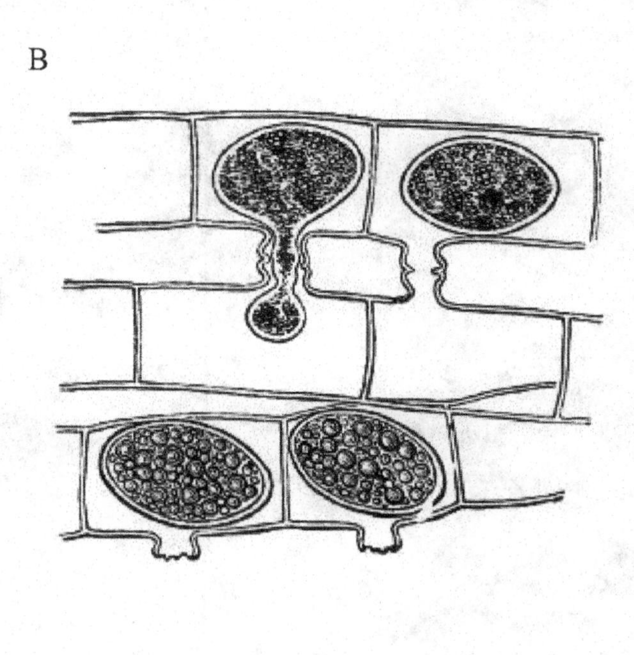

Coleochaete [Coleochaetales) is often cited as representing the type of charophycean green alga, which might have served as ancestral stock for embryophytic plants. In addition to the usual features cited as common to the charophycean green algae and plants, Coleochaete exhibits "incipient embryophytism", (i.e, there is a tendency for the zygote to be retained within the female gametangium with a layer of sterile cells enveloping the zygote fertilization).

NOTE: ***Coleochaete*** nor any algal form, is truly embryophytic!

Study the prepared slide ***Coleochaete***. Note the vegetative organization of the thallus and the enveloped zygote(s). Antheridia are very difficult to discern in this species of the genus.

Figure 5.11. Coleochaete vegetative cells

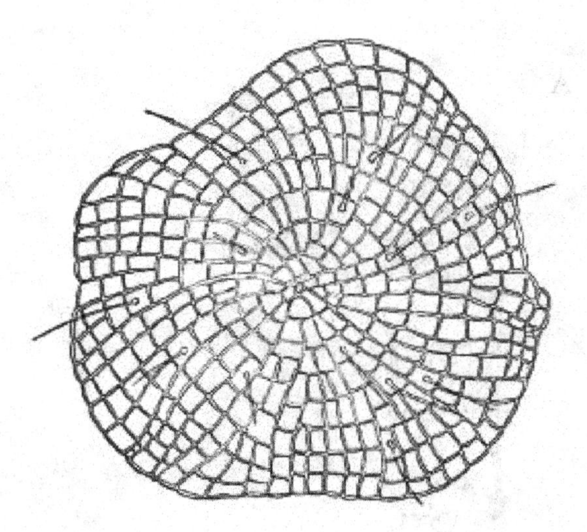

LABORATORY 6.
THE NON-VASCULAR EMBRYOPHYTES -- "BRYOPHYTES"

The Transition to Land

One of the most significant steps in the evolution of plants, an event occurring some 400 million years B.P., which was to change the face of the Earth was the adaptive radiation of photosynthetic organisms from the ancestral aquatic environment to the terrestrial environment. Although the exact nature of the ancestral form(s) is not known, it is generally agreed that the ancestral group(s) was similar to extant charophycean green algae (DIVISION CHLOROPHYTA/CLASS CHAROPHYCEAE **sensu lato**). This hypothesis is based upon a number of similarities between extant charophytes and terrestrial plants including, but not limited to, (1) common photoreceptive pigments including Chlorophyll a and b and a variety of carotenes [notably beta-carotene] and xanthophylls [notably lutein]; (2) production of true starch [I_2KI positive] as a reserve food material and its storage with in the chloroplast; (3) phragmoplastic cytokinesis and (4) common ultrastructural features, particularly chloroplast and flagellar organization.

Compared to spending one's life in an aquatic environment, maintenance in a terrestrial environment presents new challenges including, but not limited to, (1) acquisition of water and dissolved nutrients and conduction of same, (2) conduction of food and nutrients to all living cells of the plant body, (3) conservation of water, (4) gas exchange with the atmosphere, (5) mechanical support systems ["**skeleton**"] and (6) independency of the sexual reproductive system upon environmental water as a medium for gamete transmission. Descendents of the ancestral form(s) that migrated to the terrestrial environment evolved along two different lines: (1) a non-vascular plant line ["**bryophytes**"] and (2) a vascular plant line ["**tracheophytes**"]. Representatives of the first line failed to develop highly specialized conducting tissues, xylem and phloem, and mechanical support tissues. Additionally, all remain dependent upon the presence of environmental water as a medium for transmission of the flagellated sperm to the larger, non-motile egg. These plants, the "**bryophytes**", are sometimes referred to -- primarily by zoologists! -- as the "amphibians of the plant kingdom". The second line, the "**tracheophytes**", developed highly specialized conducting tissues, xylem and phloem, as well as mechanical support tissues and, ultimately, completely internalized their sexual reproductive system. The most highly evolved of the latter line, the angiosperms ["flowering plants"], today constitute the conspicuous and predominant elements of the Earth's vegetation.

KINGDOM PLANTAE -- THE "BRYOPHYTES"

Traditionally, the non-vascular, embryophytic plants, "the bryophytes", were placed into three Classes, the MUSCI (mosses), the HEPATICEAE (liverworts) and ANTHOCEROTAE (hornworts). More modern systems segregate the three groups into three Divisions, the BRYOPHYTA (mosses), the HEPATOPHYTA (liverworts) and the ANTHOCEROPHYTA (hornworts). Traditionally, the term bryophyte was used in a collective sense for all three groups. With today's classification one must be cautious when using that term, as it is often used in a more restrictive sense to mean "the mosses"! In any case, the three groups do share a number of features:

(1) They photoautotrophic; all possess Chlorophyll a as the primary photoreceptive pigment and evolve oxygen as a by-product of photosynthesis;
(2) They exhibit a heteromorphic life cycle with the gametophyte generation being vegetatively predominant; the sporophyte is permanently attached to the gametophyte and, to a great extent, nutritionally dependent upon the gametophyte;
(3) Though they achieve tissue/organ level organization, they lack specialized vascular tissues

homologous to the xylem and phloem of true vascular plants;
(4) Their reproductive structures are multicellular and enclosed within sterile cells;
(5) In sexual reproduction, they retain the zygote within the female gametangium, the **archegonium** (pl. archegonia), for development into a multicellular embryo, i.e., they are embryophytic.

In today's lab, you will be studying representatives of the non-vascular embryophytic plant line, "**bryophytes**". As you do so, consider them in light of the extent of adaptation to the terrestrial environment. What "problems" associated with life on land have they solved? What "problems" associated with life on land have they failed to solve?

DIVISION BRYOPHYTA – mosses

The Vegetative Moss Gametophyte

Obtain a living, sexually immature moss gametophyte. Using a dissecting microscope, examine it carefully. The plant has a well-defined central axis, [the **caulidium** that is attached a spirally arranged series of photosynthetic organs, the **phyllidia** leaves]. The gametophyte is attached to the substrate by rhizoids. You may not be able to see the rhizoids on the specimen; they are delicate filaments of cells that are often torn off when removing the gametophyte from its natural substrate. Note that the gametophyte is upright rather than being dorsi-ventrally flattened.

Figure 6.1. Moss gametophyte and sporophyte plant

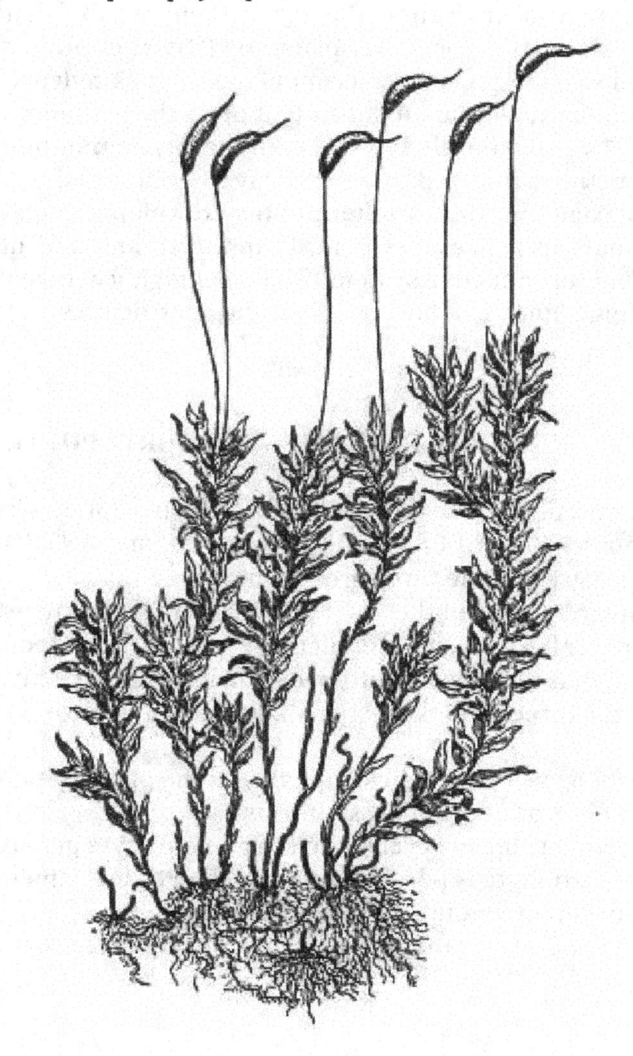

Study the prepared slide "of a cross-section of a moss" **phyllidium** (probably Mnium sp.). Note the enlarged "in the center of which lies the" **conducting strand.**

Note that, except for the region of the mid-rib, the phyllidium is only one cell in thickness. The conducting strand contains **hydroids**, water- and nutrient-conducting cells which are analogous [homologous] to the cells of vascular plants and **leptoids**, water- and food conducting cells which are analogous [homologous] to the **phloem** of vascular plants. The conducting strand of the phyllidia is continuous with that of the caulidium to which the phyllidia are attached. Sketch a portion of a phyllidium labeling: **conducting strand**

Demonstration:

Study the prepared slide of a cross-section of a moss caulidium; the slide may be labeled: "Moss Stem"

Figure 6.2. Moss caulidium cross section.

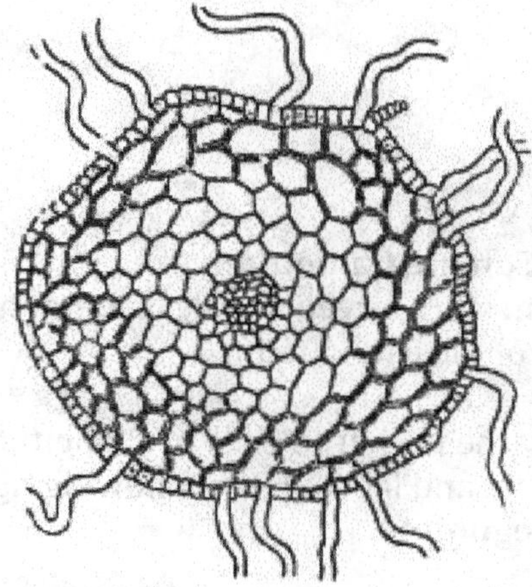

Moss Gametophyte Axis

You will note a central region consisting of relatively thick-walled cells similar to that you observed in the mid-rib of the moss phyllidium. This is the conducting strand containing two types of conducting cells, **hydroids** and **leptoids**. In cross-section it is difficult to separate these two types of cells though, in general, the hydroids tend to be larger in diameter than are the leptoids. Recall that the tissues of the conducting strand of the caulidium are continuous with those of the phyllidium. Label the following figure: **epidermis, cortex**, and **conducting strand**.

The Sexually Mature Moss Gametophyte

Upon sexual maturation of the gametophytes, the gametangia -- **archegonia** -- differentiate. The gametophytes of some mosses are **unisexual**, others produce gametophytes that are **bisexual**.

Explain the difference and indicate the advantages and disadvantages of each. The moss you are studying has unisexual gametophytes.

The Male Gametophyte

Obtain a preserved male gametophyte plant. The male gametophyte bears antheridia at the apex of the caulidium. Examine a male gametophyte plant with the dissecting microscope, noting the "**antheridial head**" at the plant's apex.

Study the prepared slide of a longitudinal section through the apex of a male gametophyte. Identify and label on the following figure: **antheridial head, antheridium**, **sperm cells**

Figure 6.3. Moss antheridial head with antheridia.

The Female Gametophyte

Obtain a preserved female gametophyte plant. The female gametophyte bears archegonia at the apex of the caulidium.

Study the prepared slide of a longitudinal section through the apex of a female gametophyte. Identify and label on the following figure: **archegonial neck**, **archegonial venter**, and **archegonium**

Figure 6.4. Moss **archegonial head and archegonium**

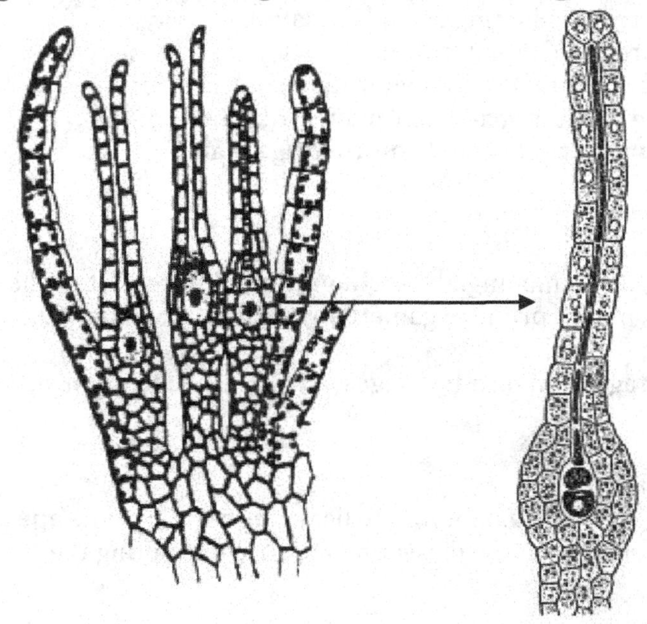

The Moss Sporophyte

Subsequent to fertilization, the zygote is retained within the archegonium of the female gametophyte. The embryo sporophyte develops and grows while remaining attached to the parental female gametophyte; the mature sporophyte, consisting of **capsule**, is largely dependent upon the parental female gametophyte for water, food and nutrients.

Obtain a preserved female gametophyte plant to which is attached a mature sporophyte. Examine the specimen with the dissecting microscope noting the and, especially, the **capsule**. If still present, note the **calyptra** covering the capsule. What is the origin of the calyptra? After initial observation, use a dissecting needle to remove carefully the **calyptra** from the capsule.

Note that the capsule has a "at its apex, the **operculum**". Carefully remove the operculum. Around the "of the exposed capsule apex you will note a ring of tooth-like cells constituting the **peristome**. The peristome represents the spore dispersal mechanism of mosses. Rupture the capsule with the dissecting needle, noting the mass of spores released. Label the following figure: **parental female gametophyte, calyptra, capsule, operculum,** and **peristome.**

Figure 6.5. Moss sporophyte stage and capsule.

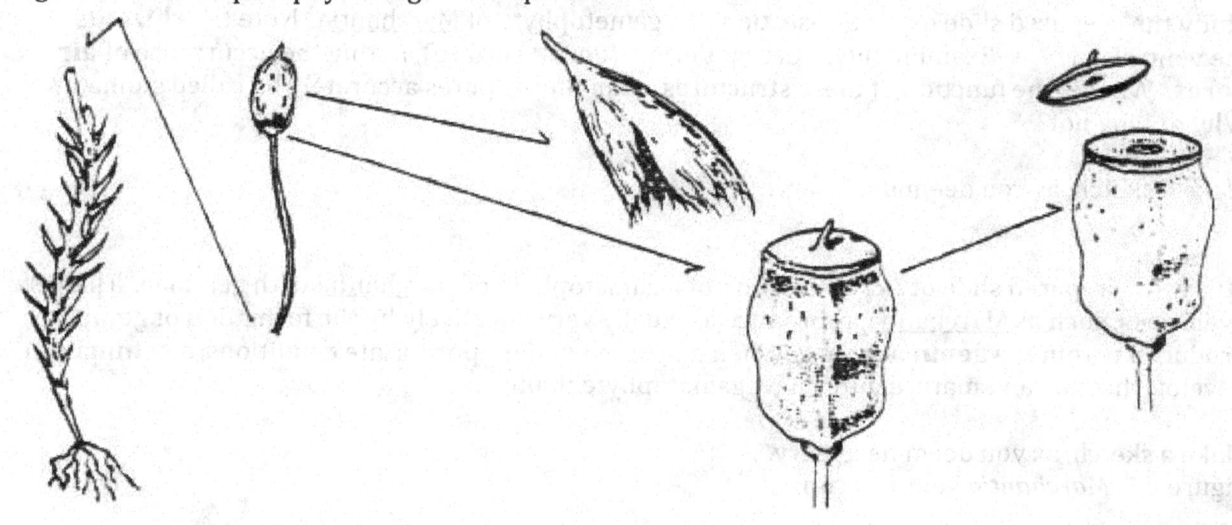

Study the prepared slide of a longitudinal section of the capsule of a moss sporophyte. Only a very small portion of the seta is included on the slide. Label the following figure: **columella** (a central column of sterile cells), **operculum, sporogenous tissue.**

Figure 6.6. Longitudinal section of a Moss capsule.

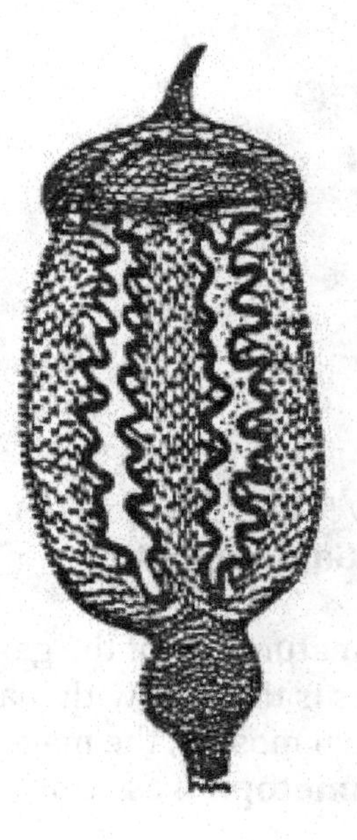

DIVISION HEPATOPHYTA – liverworts

The gametophyte of some liverworts is "**thallose**", i.e., a dorsi-ventrally flattened body that does not have stem-like and leaf-like structures; others are "in that", although dorsi-ventrally flattened, they have stem-like and leaf-like structures.

The Vegetative Liverwort Gametophyte
Demonstration:

Observe the dorsal and ventral surfaces of a sexually immature gametophyte of Marchantia, a thallose liverwort. Note the numerous **rhizoids** on the ventral surface.
Note that the thallus is ribbon-like and dichotomously lobed.

Study the prepared slide of a cross-section of a gametophyte of Marchantia. Note the **rhizoids** on the ventral surface. Examine the upper epidermis (dorsal surface), noting the occurrence of **air pores**. What is the function of these structures? Can the air pores accurately be called stomata? Why or why not?

Make a sketch, as you deem necessary.

Study the prepared slide of a cross-section of a gametophyte of Marchantia with gemmae. Thallose liverworts, such as Marchantia, reproduce asexually very effectively by the formation of **gemmae** produced within the **gemmae cups**. When dispersed under appropriate conditions, a **gemma** can develop rhizoids and mature into a new gametophyte thallus.

Make a sketch, as you deem necessary.
Figure 6.7. *Marchantia* gemmea cup.

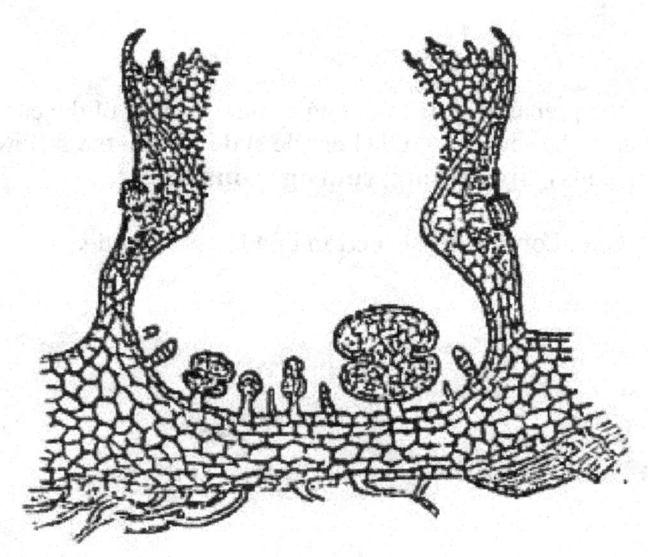

The Sexually Mature Liverwort Gametophyte
The Female Gametophyte

Upon sexual maturation of the gametophytes, the gametangia -- **archegonia** -- differentiate. Marchantia, as is the case with many liverworts, has unisexual gametophytes. As is the case with many unisexual mosses, the male and female gametophytes can be readily distinguished by their distinctive **gametophores**. Obtain a preserved female gametophyte plant Marchantia. Examine the plant with the dissecting microscope. Note that the umbrella-shaped **archegoniophores** have finger-like projections extending from the ventral surface. The ventral surfaces of the archegoniophores bear numerous **archegonia**. Label the following figure: **female gametophyte thallus, rhizoids,** and **archegoniophore.**

Study the prepared slide of a cross-section of the archegoniophore of Marchantia. Note the "typical" structure of each archegonium consisting of an elongated and a somewhat swollen **venter** in the base of which is the egg. Label the above figure: **archegoniophore, archegonium, venter** and **egg.**

Figure 6.8. *Marchantia* female gametophyte thallus.

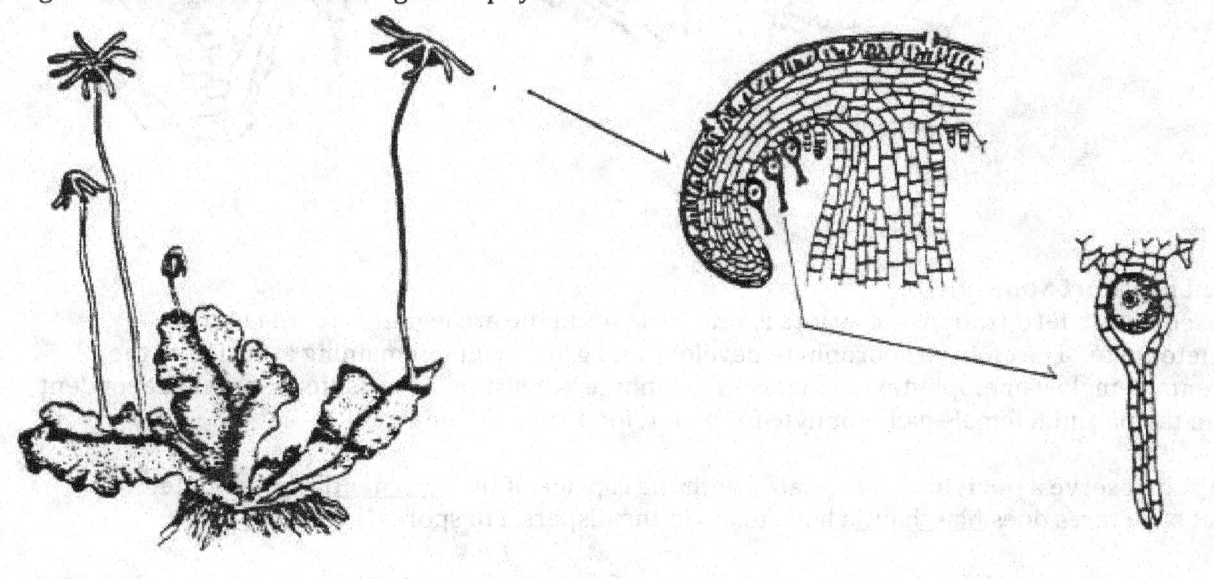

The Male Gametophyte
Obtain a preserved male gametophyte plant Marchantia. Examine the plant with the dissecting microscope. Note the disc-shaped **antheridiophores.** Label the following figure: **male gametophyte thallus, rhizoids, antheridiophore**

Study the prepared slide of a cross-section of the **antheridiophore** of Marchantia. Note that the antheridia, as are those of the mosses and all other embryophytes, are multicellular and enclosed within a sterile jacket. Upon maturation, flagellated sperm are released from the antheridia. Label the above figure: **antheridiophore, antheridium, sperm cells**

Figure 6.9 *Marchantia* male gametophyte thallus.

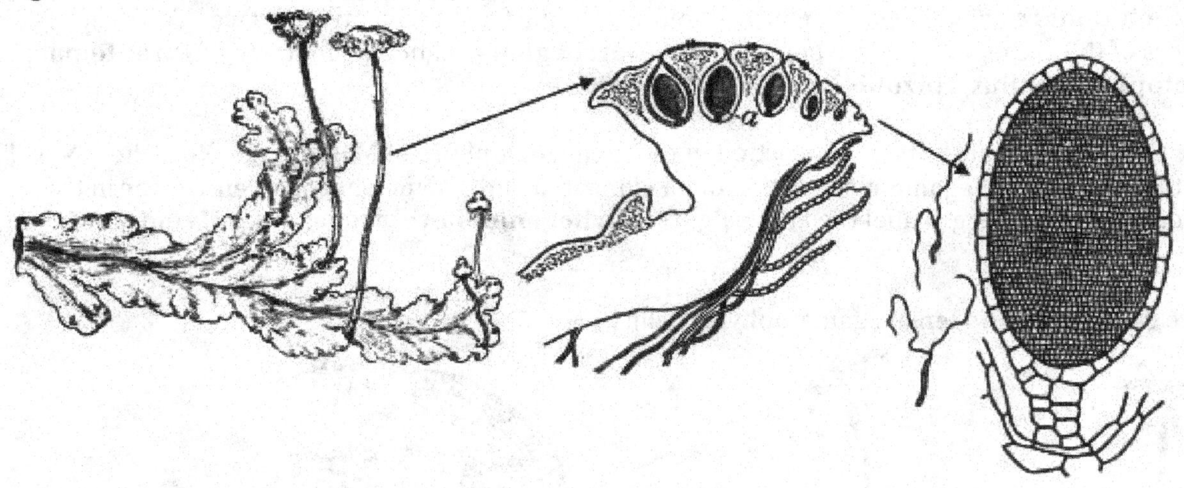

The Liverwort Sporophyte

Subsequent to fertilization, the zygote is retained within the archegonium of the female gametophyte. The embryo sporophyte develops and grows while remaining attached to the parental female gametophyte; the mature sporophyte, consisting of **capsule**, is largely dependent upon the parental female gametophyte for water, food and nutrients.

Did you observe a **peristome** associated with the capsule of the Marchantia sporophyte? If not, what structures does Marchantia have to aid in the dispersal of spores?

Study the prepared slide of a longitudinal section of a mature sporophyte of Marchantia. Label the following figure: **parental female gametophyte tissue**, **foot**, **seta**, **capsule** - within the capsule label: **elater, spores**.

Figure 6.10. *Marchantia* sporophyte capsule.

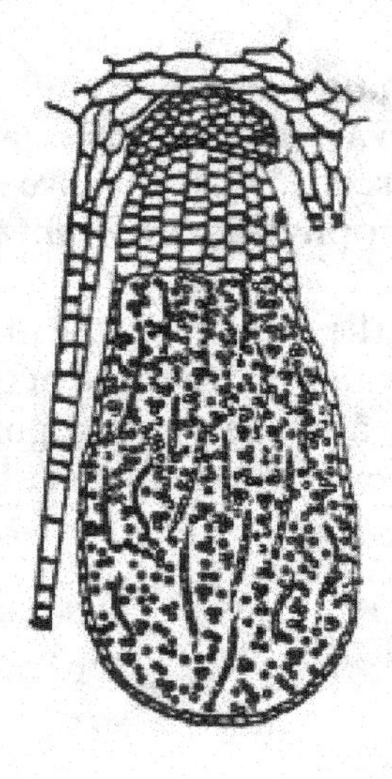

FILL IN THE APPROPRIATE SECTIONS IN THE TABLE ON PAGE 78

LABORATORY 7.
PRIMITIVE CRYPTOGAMOUS VASCULAR PLANTS

The evolution of vascular plants is inseparably related to their predominantly terrestrial habit. Their adaptation to the terrestrial environment has involved many changes including, but not limited to, the development of (1) stems, (2) leaves, (3) root systems, (4) water-proofing secretory products, (5) efficient conducting and mechanical support tissues, (6) predominance of the sporophyte generation in the heteromorphic life cycle and (7) internal sexual reproductive systems, culminating in the development of the seed habit among gymnosperms and angiosperms.

The earliest known vascular plants, whose morphology and anatomy are reasonably well known, representatives of the DIVISION RHYNIOPHYTA, date back to the mid Silurian period -- about 420 million years B.P. Fossils of a second Division of extinct cryptogamous vascular plants -- DIVISION ZOSTEROPHYLLOPHYTA -- are known from the Devonian period -- about 408-370 million years B.P. The remaining group of crypto gamous vascular plants whose representatives are all extinct, the DIVISION TRIMEROPHYTA, first appeared in the early Devonian period -- about 295 million years B.P.

The four groups of cryptogamous vascular plants with extant representatives --DIVISIONS PSILOTOPHYTA, LYCOPHYTA, SPHENOPHYTA and PTEROPHYTA -- are to be studied in this laboratory period.

DIVISION LYCOPHYTA -- lycopsids (club- "mosses" and quillworts)

Though well represented in the fossil flora, particularly of the Carboniferous period --some 360-320 million years B.P. -- this group is represented by only some fifteen extant genera. Among these, three – Lycopodium, Selaginella, and Isoetes -- occur in the flora of Arkansas. Though some extinct forms had active vascular cambia producing secondary growth and a resulting tree habit, all living forms are herbaceous. The lycopsids represent the only line of microphyllous plants with extant representatives.

Lycopodium, a homosporous lycopsid

Demonstrations:

Examine the living and herbarium specimens of the sporophyte of Lycopodium. The plant body consists of a **rhizome** that arise leafy aerial **adventitious roots**. The leaves are **microphylls** stem and roots are **protostelic**. The **sporangia** occur singly upon the upper surface of **fertile microphylls – sporophylls** – which are leaves modified to bear sporangia. The sporophylls of Lycopodium are aggregated into **strobili**, sometimes called "cones", at the apices of aerial branches. The gametophyte, though not available for your examination, is small, unisexual and, depending upon the species, is either subterranean and mycorrhizal or occurs on the substrate surface and is photosynthetic. The sperm released from the antheridia are flagellated, and water must be present as a medium for transmission of the sperm to the non-motile egg within the female gametangium, the archegonium.

Study the prepared slide of a longitudinal section of the strobilus of Lycopodium. Note that the least mature sporangia is near the apex of the strobilus; it is in these sporangia that you can observe **spore mother cells**. As you proceed "downward", you will encounter more mature sporangia in which the spore mother cells have undergone meiosis, but the spores derived from a single spore mother cell are still adherent, forming a **tetrad**. Still further "down" the strobilus you will encounter sporangia in which the tetrads have separated into individual haploid **spores**. Note that within the mature sporangia, all spores are essentially identical in size. This condition where spores are of one type and give rise to either unisexual or bisexual gametophytes is called **homospory**. Label the following figures: **strobilus, microphyll, sporophyll**, and **sporangium**.

Figure 7.1. *Lycopodium* vegetative plant and strobilus.

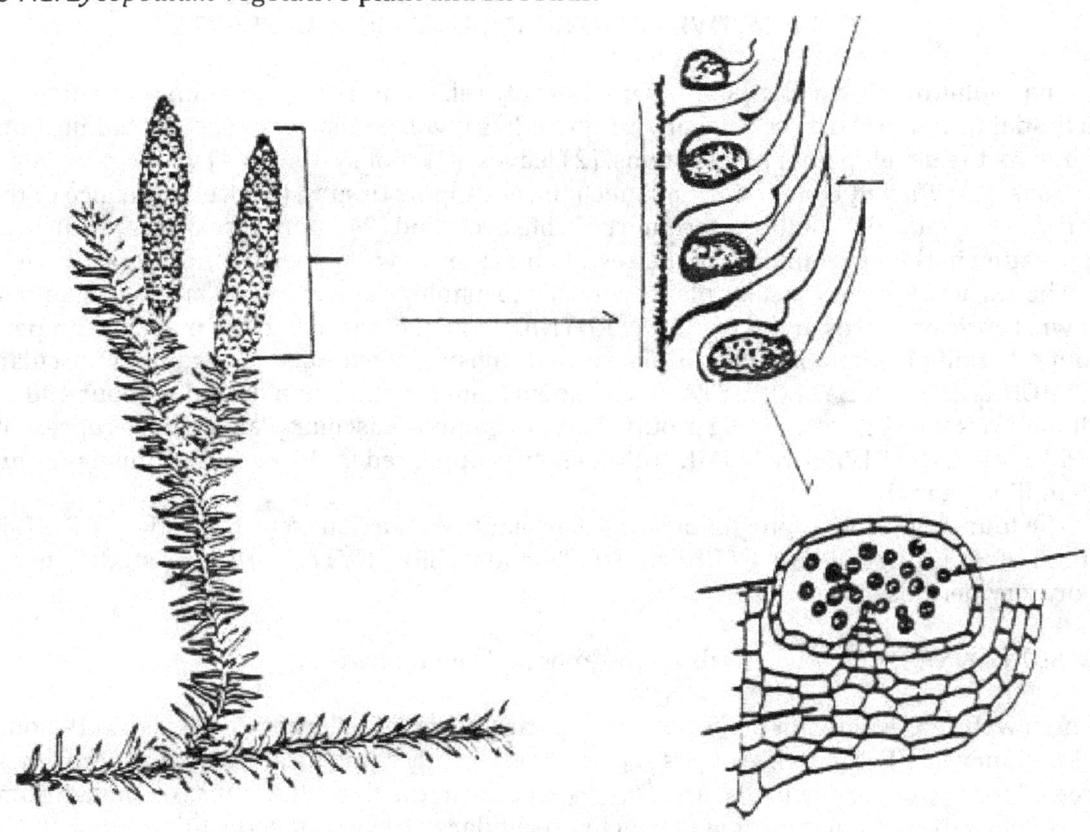

Selaginella, a heterosporous lycopsid

The herbaceous sporophyte of Selaginella is in many respects similar to that of Lycopodium. They differ, however, in a very important respect; species of Selaginella **heterosporous**. Each sporophyll within a strobilus bears a single sporangium on its upper surface. Being heterosporous, the larger of the two types of spores produced, the **megaspore**, is produced in limited number -- typically four --within a **megasporangium** borne on the surface of a **megasporophyll**. Upon germination, a megaspore will produce a female gametophyte (**megagametophyte**). Conversely, the smaller of the two types of spores, the **microspore**, is produced in large numbers within a **microsporangium** borne on the surface of a **microsporophyll**. Upon germination, a microspore will produce a male gametophyte (**microgametophyte**). Thus the gametophytes of Selaginella, and all heterosporous forms, are unisexual. The megagametophyte, upon maturation, produces **archegonia**; the microgametophyte, upon maturation, produces **antheridia**. The sperm produced by the antheridia are flagellated, and water must be present for transmission of the sperm to the non-motile egg in the archegonium. Both the micro- and megagametophyte essentially complete their development and maturation while retained within the micro- and megaspore walls, respectively; i.e., their development is endosporic. The appearance of heterospory and endosporic development of the micro- and megagametophyte was an important event in the evolutionary history of vascular plants as they "set the stage" for development of the seed habit. All seed plants are heterosporous.

Demonstrations:

Examine the living and herbarium specimens of the sporophyte of Selaginella. The plant body consists of a stem from which arise the leaves – **phylls** and **adventitious roots**. The stem and root are **protostelic**. The sporangia are aggregated into **strobili**, sometimes called "cones". Since Selaginella is heterosporous, the strobili contain two types of fertile microphylls: **microsporophylls** each bearing a single **microsporangium** and **megasporophylls** each bearing a single **megasporan**. The micro- and megagametophytes of Selaginella are not available for your examination.

Study the prepared slide of a longitudinal section of the strobilus of Selaginella. Note that each **microsporophyll** bears a single **microsporangium**, which subsequent to meiosis and maturation, contains numerous **microspores**. Conversely, **megasporophyll** bears a single **megasporangium**, which subsequent to meiosis and maturation contains four **megaspores**. Label the following figure: **strobilus, microsporophyll, microsporangium, microspore, megasporophyll, megasporangium**, and **megaspore**.

Figure 7.2. *Selaginella* vegetative plant and strobilus.

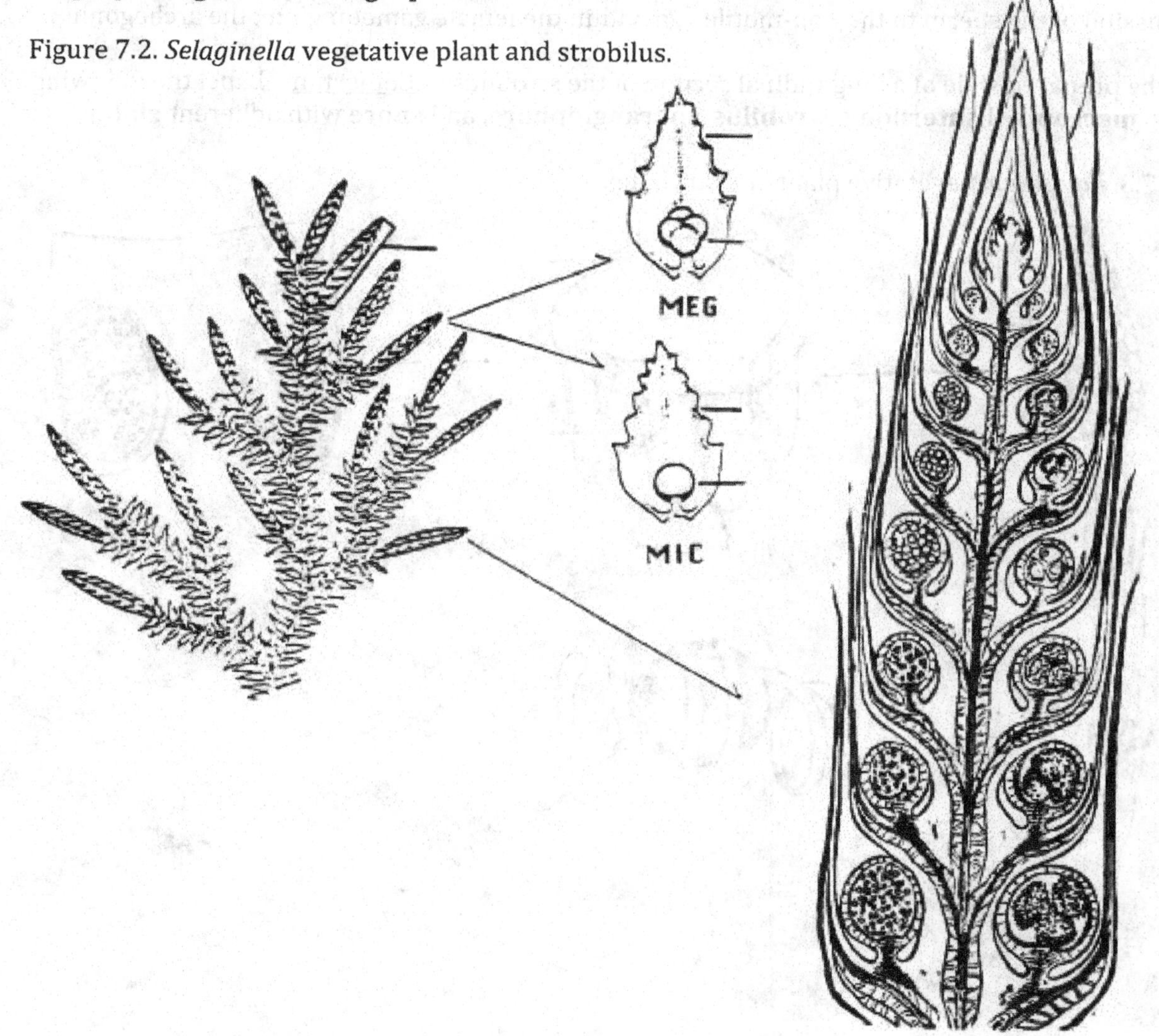

DIVISION SPHENOPHYTA -- sphenopsids (horsetails; scouring rushes)

As the lycopsids, the sphenopsids are well represented in the fossil flora, particularly of the Devonian and Carboniferous periods. This group, however, is represented by only one extant genus, Equisetum. Some of the extinct forms, particularly those of the Carboniferous period, had active vascular cambia producing secondary growth and a resulting tree habit. All species of Equisetum are herbaceous.

Demonstrations:

Examine the living and herbarium specimens of the sporophyte of ***Equisetum***. The plant body is differentiated into **aerial stem, leaves** (macrophylls), **rhizome**, and **adventitious roots**. Note that the macrophylls are borne in whorls upon the stem; those points on a stem from which arise lateral branches or leaves are called **macrophylls**. The region of a stem between two adjacent nodes is called **internode**. The **sporangia** are borne in groups of 5-10 circumferentially arranged upon the lower surface of a unique, umbrella-shaped structure, the **sporangiophore**, which is a modified branch. You will recall from previous study that the sporangia of the lycopsids are borne on the upper surface of modified leaves, the sporophylls. Note that the sporangiophores of ***Equisetum*** are aggregated into **strobili** stem apices. The gametophytes, though not available for your examination, are small, bisexual and chlorophyllous. The sperm released from the antheridia are flagellated, and water must be present as a medium for transmission of the sperm to the non-motile egg within the female gametangium, the archegonium.

Study the prepared slide of a longitudinal section of the strobilus of Equisetum. Label the following figures: **macrophyll, internode, strobilus, sporangiophore**, and **spore** with adherent **elater**.

Figure 7.3. *Equisetum* vegetative plant and strobilus.

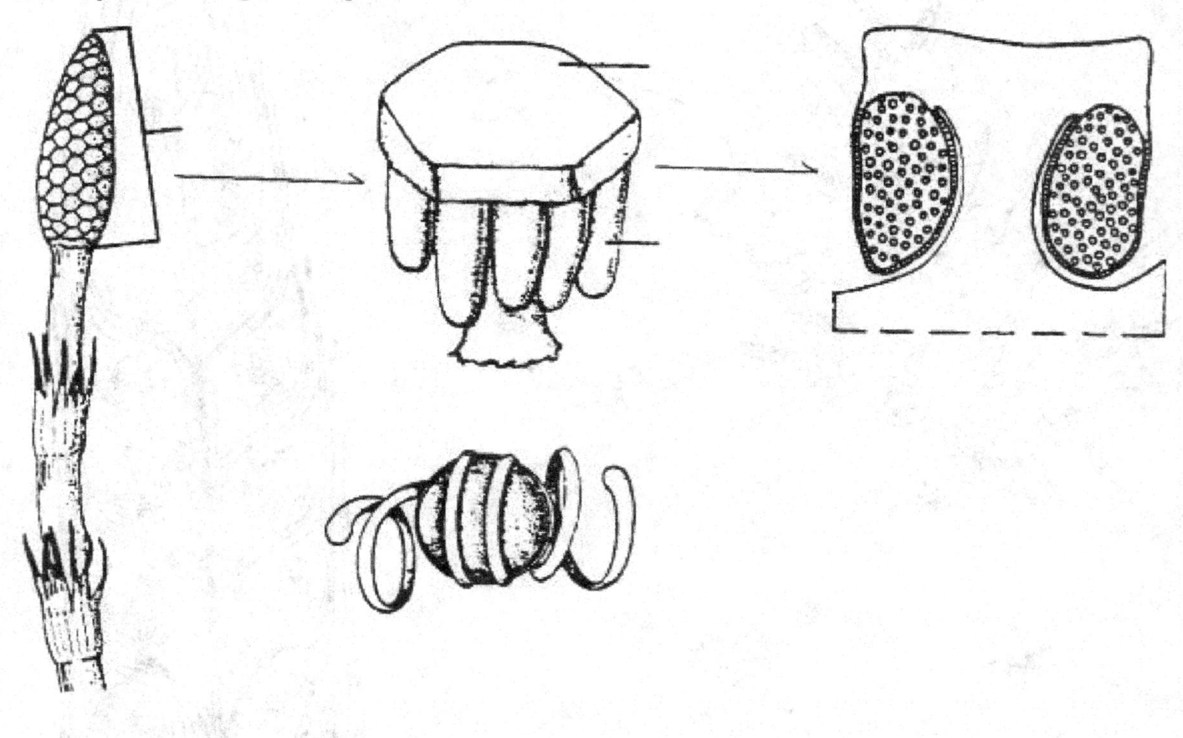

Study the prepared slide of a cross-section of the aerial stem of ***Equisetum***. Note that ***Equisetum*** has a siphonostele. The center of the stem has a large air cavity where, in earlier stages of development, occurred the pith; the parenchyma cells of the pith break down, leaving the stem hollow. Note that just external to the central air cavity (pith) there occurs a radially arranged series of vascular strands adjacent to small air cavities; the latter are called **carinal canals**. Each carinal canal is associated with a strand of **phloem**. You will note that there is a radially arranged series of air cavities in the **cortex** as well. Sketch a wedge-shaped portion of the stem cross-section and label: **epidermis, cortex, carinal canal**, and **phloem.**

DIVISION PTEROPHYTA -- pteropsids (ferns)

The ferns appear relatively abundantly in the fossil record back to the Carboniferous period. Among the cryptogamous vascular plants, the ferns are represented in today's flora in rather great diversity; indeed, in respect to the number of extant species, they are second only to the angiosperms. The leaves, sometimes called **fronds**, are macrophylls and represent the most conspicuous part of the sporophytic plant body. The ferns are the only cryptogamic vascular plants that possess well-developed macrophylls.

Demonstrations:

Study the living and herbarium specimens of fern sporophytes. Note that the leaves (macrophylls) are large and **compound** i.e., the leaf lamina is divided into leaflets that are attached to the **rachis**, an extension of the petiole. The leaves arise from the **rhizome** that may bear numerous **adventitious roots**. Note that the young, emerging leaves are coiled (circinnate); they are commonly referred to as "fiddleheads". Note on the lower surface of the leaves. Label the following figure: **macrophyll** (compound), **fiddlehead**, **rhizome**, **adventitious roots**.

Figure 7.4. Fern vegetative plant.

In respect to the position of sporangia on the sporophyll, how do these ferns differ from all other vascular plants heretofore studied in the lab?

Figure 7.5. Fern pinna with sori.

Study the prepared slide of a cross-section of a fern **sorus**. Label the following figures: **macrophyll, sporangium, annulus**, and **lip cells**.

Figure 7.6. Cross section of a sorus with sporangia.

Study the prepared slide of a whole mount of a fern gametophyte (prothallus). The archegonia tend to be aggregated near the "cleft" of the prothallus; the antheridia tend to be aggregated at the "end" of the prothallus in the area bearing rhizoids. If the slide you are using has unisexual gametophytes, you will need to study both the male and female gametophytes. Label the following figure: **vegetative body of gametophyte**, **rhizoid**(s), **antheridium**, and **archegonium**.

Figure 7.7. Fern prothallus with young sporophyte plant.

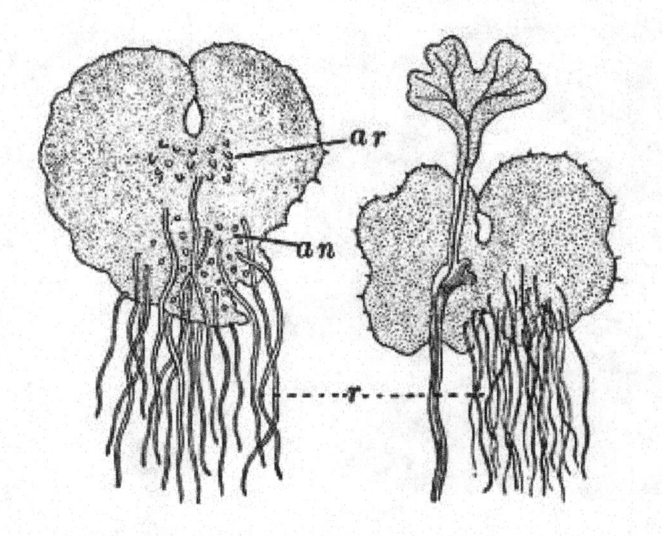

Demonstration:

Study the prepared slide of a whole mount of a fern gametophyte to which is attached a young sporophyte. Label the following figure: **vegetative body of gametophyte**, **primary root of sporophyte**, and **foliage leaf of sporophyte**.

FILL IN THE APPROPRIATE SECTIONS IN THE TABLE ON PAGE 78

LABORATORY 8.
INTRODUCTION TO PHANEROGAMOUS PLANTS
THE CONE-BEARING SEED PLANTS

The remaining groups of vascular plants are phanerogamous; i.e., they produce seeds as a result of their sexual reproductive process. In these plants, the male gametophyte (microgametophyte) is greatly reduced in size and produced within the wall of the microspore; i.e., its development is endosporic. A microspore and its enclosed mature microgametophyte are commonly called a **pollen grain**. Thus, a "delivery system" developed for transmitting the sperm to the egg while the former, as well as the latter, is permanently protected from the hostile, dessicatory nature of the terrestrial environment. Nevertheless, "old habits die hard" such that flagellated sperm are still produced by the more primitive seed plants. The dependency upon water in the habitat to effect transmission of the sperm to the egg -- and the last link to the ancestral aquatic environment --was broken at last! Concomitant with this change in the male gametophyte, the female gametophyte (megagametophyte) also underwent dramatic change with its enclosure and retention within the megasporangium wall (**nucellus**) and, additionally, within maternal sporophytic tissue, the **integuments**. The nucellus, integuments and enclosed megagametophyte comprise a structure called the **ovule**. It is the ovule, after fertilization of the enclosed egg, which becomes what we commonly call a **seed**. The old-fashioned and now out-moded archegonium has been retained by the gymnospermous seed plants but is lacking in the female gametophyte of the angiospermous seed plants. All seed plants are siphonostelic and macrophyllous. One can recognize two major subgroups of seed plants, the "gymnosperms" and the "angiosperms".

Among the former there are four Divisions with extant representatives: CYCADOPHYTA, GINKGOPHYTA, GNETOPHYTA and CONIFEROPHYTA. The gymnosperms have "Naked" seeds; i.e., the seeds are not enclosed within a fruit derived from maturation of an ovary and, occasionally, other associated structures. All angiosperms, commonly called "flowering plants", are placed into a single Division, the ANTHOPHYTA, and have their seeds enclosed within a fruit.

The Gymnospermous Seed Plants

The gymnosperms are well represented in the fossil flora of the Carboniferous period but achieved dominance during the Triassic. The plant body of the gymnosperms is differentiated into stems, leaves and roots. The leaves are macrophyllous and, in many cases, "evergreen" rather than deciduous. They typically produce pollen in association with microsporangiate strobili ("pollen cones") and ovules in association with megasporangiate strobili ("ovulate cones"). Secondary growth, due primarily to the activity of a vascular cambium, is of common occurrence. This is particularly true in the more advanced conifers, and, as a result, many gymnosperms are trees.

DIVISION CONIFEROPHYTA – conifers

The Earth's vegetation was dominated from this group, particularly during the Jurassic into the Cretaceous periods. They first appear in the fossil record during the Carboniferous. Approximately 500 species survive in today's flora. The most familiar of these are the pines (***Pinus***), junipers (***Juniperus***), hemlocks (***Tsuga***), yews (***Taxus***), redwoods (***Sequoiadendron***), firs (***Abies***) and spruces (***Picea***). All living conifers, as well as some angiosperms, exhibit secondary growth due largely to the activity of the vascular cambium, a type of lateral meristem.

The genus ***Pinus*** is the largest of the extant conifers. Pines have needle-like leaves, which are typically produced in clusters called fascicles.

The Leaf
Although it is common to think of conifers, including pines, as having "evergreen" leaves and angiosperms as having deciduous leaves, some conifers have deciduous leaves, i.e., bald cypress (Taxodium), and some angiosperms have "evergreen" leaves, i.e., **magnolia**.

Study the prepared slide of a cross-section of a leaf. Note that epidermis has a very thick **cuticle**. The **stomata** and their bordering **guard cells** sunken below the surface.

Internal to the epidermis appears one (or more) layer(s) of closely packed, very thick-walled cells comprising the **hypodermis**.

Just internal to the hypodermis, there is a multi-celled layer of **parenchyma** constituting the **mesophyll**. The parenchyma cells of the mesophyll are easily recognized as their cell wall margins are **crenulated**, and the cells fit together closely as the pieces of a jigsaw puzzle. Note the **resin ducts** in the mesophyll. The conducting tissues of the leaf are separated from the mesophyll by the **endodermis**, a single layer of cells. Just internal to the endodermis lies **transfusion tissue**.

Depending upon the species, the leaf will have one or two vascular bundles within the transfusion tissue. The vascular bundle(s) contains the **xylem tracheids** and the **phloem**. Sketch a portion of the leaf labeling: **epidermis, guard cell, hypodermis, mesophyll, endodermis, transfusion tissue, xylem**, and **phloem**.

Figure 8.1. Cross section of a pine needle.

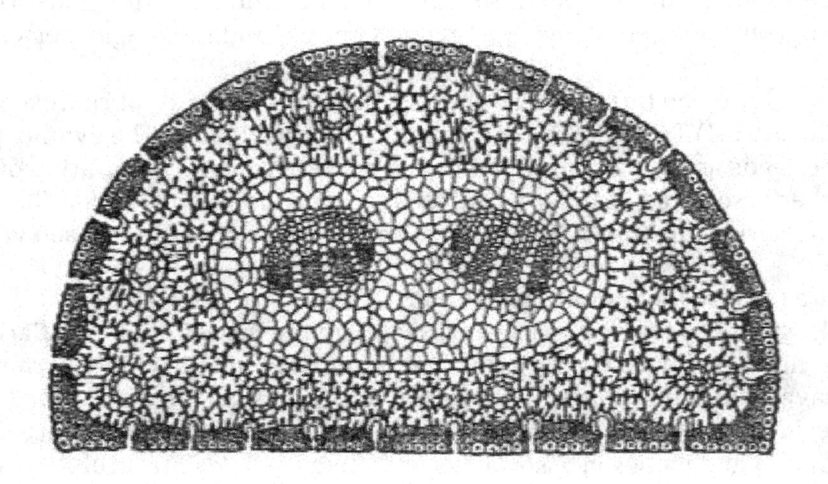

The Life Cycle of Pine

You will recall that all seed plants are **heterosporous**. The microsporangia and megasporangia of pines and other conifers are borne in separate strobili on the same plant; i.e., the plant is **monoecious**.

The Microsporangiate Strobilus ("pollen cone") and Development of the Microgametophyte

Demonstrations:
Examine the various stages of maturation of microsporangiate strobili (**pollen cones**) available in the laboratory.

Study the prepared slide of a longitudinal section of a microsporangiate strobilus (**"pollen"**). Note: The slide may be labeled, "Pine Staminate". Each **microsporophyll** bears two **microsporangia** on its lower surface; on the slide you will see only one of the two microsporangia. Sketch below a portion of the microsporangiate strobilus and label: **microsporophyll, microsporangium, wings, microspores, microspore mother cells** (i.e., if meiosis has not yet occurred).

Figure 8.2. (A) Male pine cones (B) female pine cone (C) cone scale (D) pollen grain.

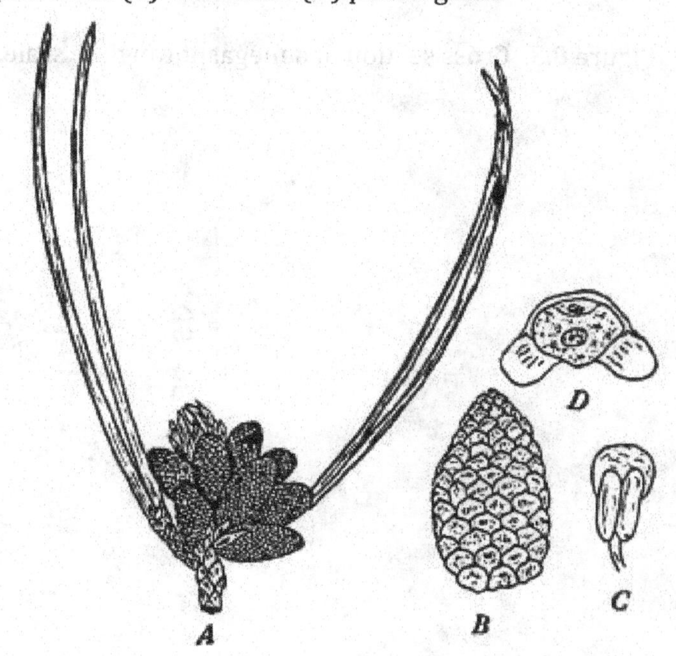

Continue your study of the prepared slide of the microsporangiate strobilus. Carefully, examine the microspores within a mature microsporangium. Note that each microspore bears two bladder-shaped **wings** that aid in wind dispersal. You should be able to find a microspore with two nuclei; one of these is the **generative nucleus**, the other is the **tube nucleus**. The **prothallial cells**, which represent remnants of the vegetative body of the male gametophyte (microgametophyte) and have no role in the reproductive process, collapse shortly after formation and are often difficult to see. Note that the development of the microgametophyte is **endosporic**, a feature characteristic of all heterosporous plants. Though you will not be able to see additional development of the microgametophyte on the slide, ultimately the **generative nucleus** give rise to a **spermatogenous cell,** which ultimately produces the **sperm cell**. The **tube nucleus** gives rise to the **pollen tube**. Sketch below a single microspore labeling: **microspore wall, generative nucleus**, and **tube nucleus**.

The Megasporangiate Strobilus ("ovulate cone") and Development of the Megagametophyte

Demonstration:

Examine the various stages of maturation of megasporangiate strobili "ovulate cones" available in the laboratory. In mature cones, note the presence of depressions on the surface of an **ovuliferous scale** marking the original location of the ovules and the seeds (post-fertilization ovules). Note that the ovuliferous scales are spirally arranged on the central axis of the **megastrobilus**. Study the prepared slide of a longitudinal section of a megasporangiate strobilus.

Note: The slide may be labeled, "Pine Young Ovulate Cone". Note that you can see only one of the two **ovules** borne on the surface of each **ovuliferous scale** [ovuliferous scale + ovuliferous bract = seed scale complex]. Sketch below a portion of the megasporangiate strobilus and label: **ovuliferous scale** and **ovuliferous bract**.

Continue your study of the longitudinal section of a megasporangiate strobilus. Examine a single **ovuliferous scale/bract**. Sketch below and label: **ovule, ovuliferous scale, ovuliferous bract, integument, micropyle,** and **megaspore mother cell**.

Figure 8.3. Cross section of a megasporangiate scale.

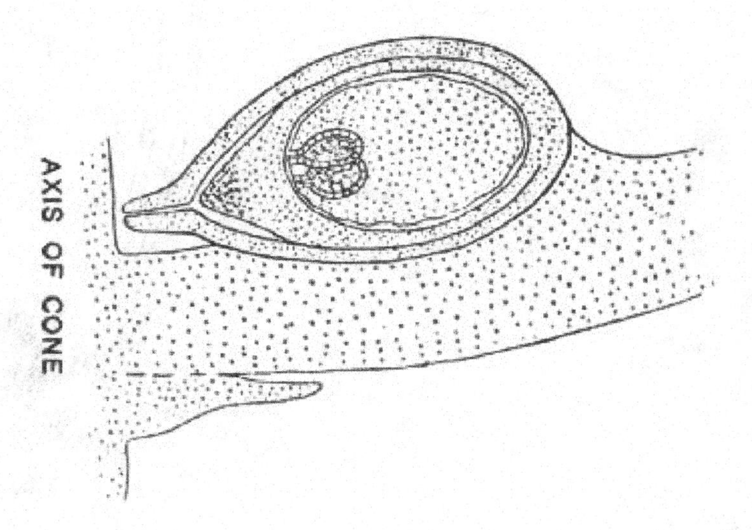

As development of the megagametophyte continues, the megaspore cell -- within the megasporangium, within the ovule -- undergoes meiosis. Of the four haploid products, three disintegrate leaving a single **megaspore** within the megasporangium. Through a succession of mitoses followed later by cytokineses, the megaspore gives rise to the **megagametophyte** that bears, at the micropylar end of the ovule, two or three **archegonia**. Each archegonium contains a single egg.

Demonstration:

Study the prepared slide of a mature megagametophyte. Sketch below and label: **integument, nucellus, vegetative body of female gametophyte,** and **archegonium**.

Figure 8.4. mature megagametophyte and the development of the gametophyte.

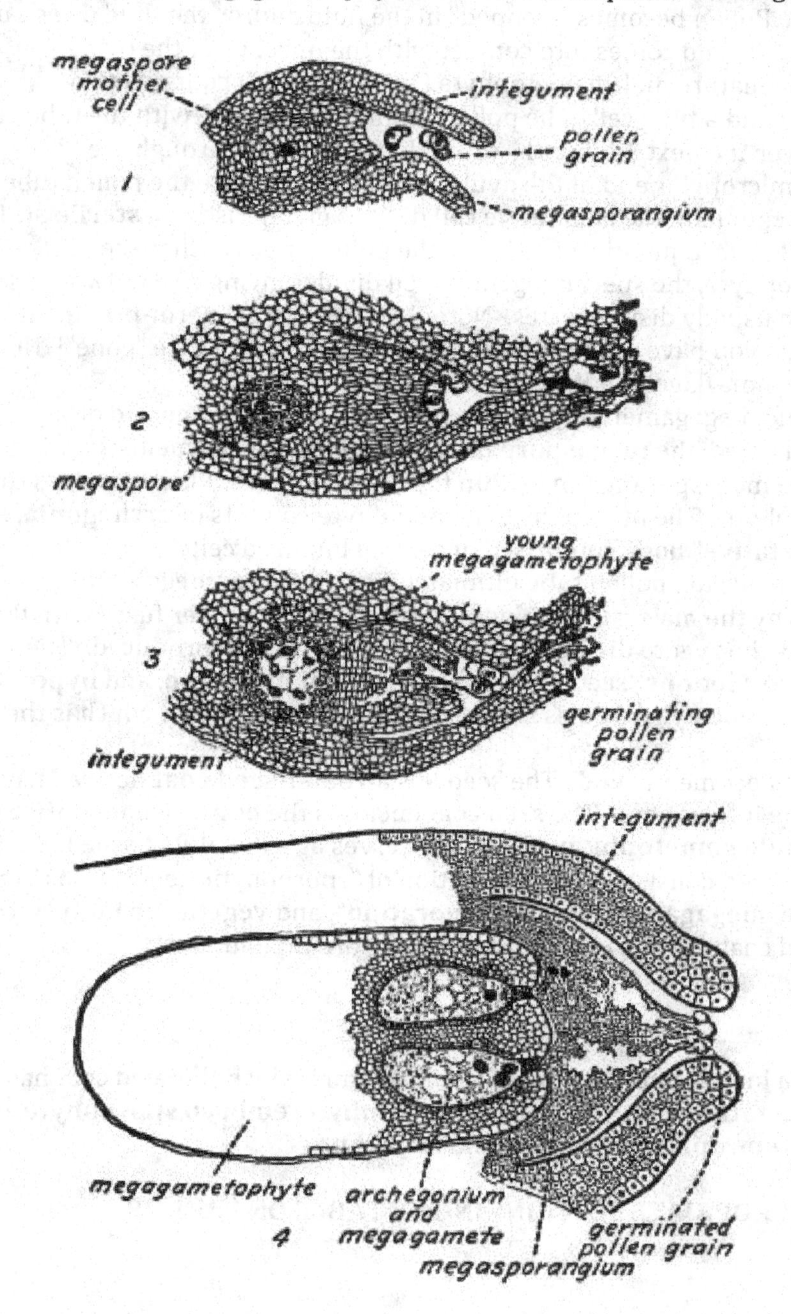

Pollination and Fertilization

Pollination is the physical transfer of pollen grains from the microsporangia of a microsporangiate strobilus to the ovuliferous scales of a megasporangiate strobilus. This process occurs in early spring. At this time, the ovuliferous scales of the megastrobilus are separated from one another ("opening"). The wind-dispersed pollen, which is produced in enormous abundance, "sifts" down into the megastrobilus. During this time, a small drop of viscous fluid exudes from the micropylar region of each ovule. Pollen becomes "trapped" in the fluid and, as the fluid dries out, pollen is "drawn" into the micropyle and comes into contact with the nucellus of the ovule.

At the time of pollination, the mature male gametophyte ("pollen grain") contains two prothallial cells, a generative cell and a tube cell. The pollen grain "germinates" with the tube cell forming a **pollen tube**, which over the next 12 months or so, digests its way through the integument and nucellus at the micropylar end of the ovule. Usually just before the pollen tube completes penetration of the integument, the generative cell divides giving rise to a **sterile stalk cell** and a **spermatogenous cell**. Subsequently, just before the pollen tube reaches the archegonium of the megagametophyte, the spermatogenous cell divides giving rise to two **non-flagellated sperm**, one of which usually disintegrates. Note that the type of sperm-producing structure, the antheridium, which you have seen repeatedly in previous groups are "gone"! Note also that the sperm of the pine is non-flagellated!

Subsequent to pollination, the megagametophyte goes through a prolonged period of development. As previously indicated, the **megaspore mother cell** undergoes meiosis, giving rise to four **megaspores** --within the megasporangium, within the ovule -- one of which survives to develop into the female gametophyte. The mature megagametophyte consists of **archegonia**, each containing an egg, and the "**vegetative**" body comprised of several hundred cells.

Upon penetration of the nucellus, the pollen tube ultimately reaches an archegonium. One of the two sperm produced by the male gametophyte disintegrates, the other fuses with the egg, giving rise to the **zygote**. As the zygote divides, the **embryo sporophyte** produced. The embryo sporophyte consists of **cotyledons** "seed leaves", **shoot apical meristem**, and **hypocotyl-root axis**. The distal end of the hypocotyl-root axis is covered by **root cap** behind which is the **root apical meristem**.

Each post-fertilization ovule becomes a **seed**. The seed has an outer **seed coat** derived from remnants of the nucellus plus the integument. The seed coat encloses the **embryo sporophyte** and the **vegetative body of the female gametophyte**. The latter serves as a nutritive tissue for the embryo sporophyte. Thus, each seed consists of a combination of "sporophytic generations", the **seed coat**, **embryo sporophyte**, **megagametophytic "generation"**, and **vegetative body of the female gametophyte**. The seed coat and the embryo sporophyte are diploid; the megagametophyte is haploid.

Demonstration:

Study the prepared slide of a longitudinal section of a seed of from which the seed coat has been removed. Sketch and label: **vegetative body of female gametophyte**, **embryo sporophyte**, **cotyledon**, **shoot apical meristem**, **apical meristem**, and **root cap**.

FILL IN THE APPROPRIATE SECTIONS IN THE TABLE ON PAGE 78

LABORATORY 9.
THE ANGIOSPERM SEXUAL CYCLE: FLOWERS, SEEDS AND FRUITS

The angiosperms -- "flowering plants" -- represent the dominant plant life of the present geological epoch except in a few biomes where conifers predominate. There are some 260,000 described species, the largest of all extant plant groups. The earliest record of the angiosperms dates back to the Jurassic. By the Paleocene, nearly all families represented on Earth today had already appeared. One, among several, of the important factors responsible for their success as terrestrial organisms is their sexual reproductive system involving flowers, seeds and fruits.

DIVISION ANTHOPHYTA
CLASS MONOCOTYLEDONAE -- monocotyledonous angiosperms (monocots)
CLASS DICOTYLEDONAE -- dicotyledonous angiosperms (dicots)

The Angiosperm Flower

The typical angiosperm flower is a composite of fertile and sterile modified leaves borne at the apex of a stem, the **receptacle**. As with any other stem, the receptacle has **internodes**. However, in the flower, the internodal distances are greatly reduced ("shortened"). As a consequence, the nodes in a flower are very close together. A flower may have, at most, four different components. Two of these, the sepals and the petals, are sterile and are attached to the receptacle below the fertile components, **stamens** and the **carpel(s)** [= pistil(s)]. These four floral components are arranged outside to inside – sepals, petals, stamens, and carpel(s).

All sepals, collectively, comprise the **calyx**; all petals, collectively, comprise the **corolla**. Together, the calyx and the corolla constitute the **perianth**. Though exceptions occur, the sepals are usually green in color, and the petals are brightly colored. The stamens are **microsporophylls**, each typically consisting of a slender attachment stalk, the **filament**, at the apex of which occurs a bi-lobed **anther**, which contains four **microsporangia** (= pollen sacs). All the stamens, collectively, constitute the **androecium**.

Figure 9.1. Flower cross section.

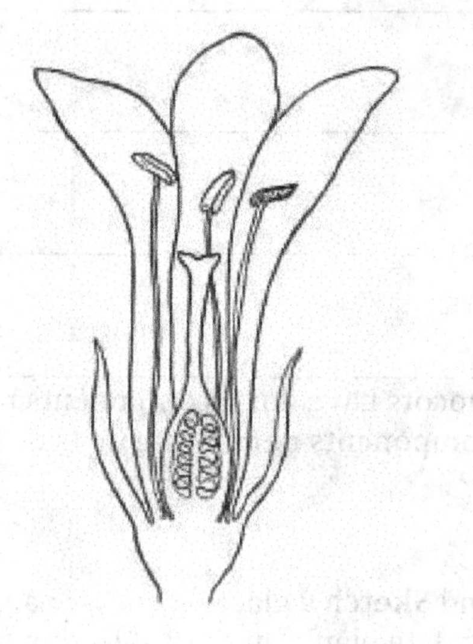

The carpels -- sometimes called pistils -- are **megasporophylls** that are enfolded lengthwise and enclose one or more **ovules**. A given flower may have one or more carpels. All the carpels of a flower constitute the **gynoecium**.

In most flowers, the gynoecium is differentiated into a swollen base, the **ovary**, which contains the ovules, and terminates in a pollen-receptive surface, the **stigma**. The stalk like portion between the ovary and the stigma is the **style**. That portion of the ovary to which the ovule(s) is(are) attached is the **placenta**.

Among the some 260,000 described angiosperm species, basic flower structure varies. For example, most flowers possess both stamens and carpels and are said to **perfect**. If either stamens or carpel(s) are lacking, the flower is said to be **imperfect**. In that case, depending upon which component is represented, the flower is either **staminate** (stamens present; carpel(s) lacking) or **carpellate** (pistillate) (carpel(s) present; stamens lacking). In the case of imperfect flowers, an individual plant may possess both types of flowers in which case the plant is said to be **monoecious**; if the individual plants possesses either staminate or carpellate flowers, the plant is said to be **dioecious**. Additionally, any one of the four floral components -- sepals, petals, stamens or carpel(s) - may be lacking in the flowers of some angiosperms. If all four are present, the flower is said to be **complete**. If one of the four is lacking, the flower is said to be **incomplete**. Thus, imperfect flowers are also incomplete! However, not all incomplete flowers are imperfect!

Another variation of floral morphology is related to symmetry. When the corolla is composed of petals of similar shape that radiate from the center and are equidistant from one another, the flower is radially symmetrical and is said to be **regular** (actinomorphic). In other cases, one or more members of the calyx <u>and/or</u> corolla are shaped differently from other members of the same whorl; such flowers are bilaterally symmetrical and are said to be **irregular** (zygomorphic).

Obtain a snapdragon flower. Observe the "gross" morphology of the flower, filling in the appropriate blanks in the following Table. Carefully dissect the flower. With a razor blade, cut a cross-section through the ovary and observe with a dissecting microscope.

Table 9.1

FLOWER	Complete/ Incomplete?	Perfect/ Imperfect?	Regular/ Irregular?	# of Stamens?	# of Petals?	# of Sepals	Monocot/ Dicot?
SNAPDRAGON							
GLADIOLUS							
Other? (Identify)							
Other? (Identify)							
Other? (Identify)							

*NOTE: Monocots have floral components numbering in 3s or multiples thereof, rarely >9; dicots have floral components numbering in 4s or 5s or multiples thereof.

Obtain and **Sketch** a gladiolus flower and observe the "gross" morphology of the flower, filling in the appropriate blanks in the Table above. Carefully dissect the flower. With a razor blade, cut a cross-section through the ovary and observe with a dissecting microscope. Label: **receptacle, stamens, anther, filament, carpel, petal, sepal, calyx, corolla, perianth, ovary, stigma, and style.**

If available, follow the same procedure for additional flowers as for the flowers of the snapdragon and gladiolus, filling the appropriate blanks in the Table above.

The Angiosperm Sexual Life Cycle

The micro- and megagametophytes of the angiosperms are reduced in size to a greater extent than any other heterosporous plant, including the gymnosperms. The mature male gametophyte consists of only three cells (nuclei), which develop within the microspore. The mature female gametophyte, permanently enclosed within parental sporophytic tissue, consists of seven cells (8 nuclei), even fewer in some angiosperms. Neither antheridia nor archegonia are present! Subsequent to pollination, the pollen tube of the microgametophyte conveys two non-flagellated sperm to the egg within the megasporangium of the ovule. After **syngamy**, the ovule develops into a seed. Concomitantly, the ovary of the carpel -- plus, in some cases, additional structures -- develops into a fruit.

Microsporogenesis and Development of the Microgametophyte

Microsporogenesis is the development of microspores within the microsporangia of the anther.

Study the prepared slide of a cross-section through an immature (pre-meiotic) **anther** of Lilium. Note the strand of vascular tissue that represents the plane of section through the **filament** of the **stamen**. You will observe that the anther is bi-lobed with each lobe containing two **microsporangia**. A layer of nutritive cells, the tapetum, that supply food to the developing microspores, internally lines each microsporangium. Internal to the tapetum are numerous **microspore mother cells** (microsporocytes). Sketch and label: **vascular strand of filament**, **microsporangium**, **microspore mother cells**, and **tapetum**.

Study the prepared slide of a cross-section through a mature (post-meiotic) **anther** of Lilium. Each microspore mother cell has undergone meiosis, giving rise to four haploid **microspores**. Using high power, find and observe a microspore with an enclosed 2-celled (binucleate) developmental stage of the male gametophyte. The cell with the spherical nucleus is the **tube cell**; the cell with the spindle-shaped nucleus is the **generative cell**. In most angiosperms, the microspores are shed from the anther in this stage of development and are commonly referred to as **pollen grains**. Sketch a pollen grain with the enclosed 2-celled male gametophyte and label: **microspore wall**, **tube cell** (nucleus), and **generative cell** (nucleus).

Demonstration:

Study the prepared slide on which pollen grains are of a variety of angiosperms. The shape, size and ornamentation of the pollen grain are species specific.

Megasporogenesis and Development of the Megagametophyte

Megasporogenesis is the development of a megaspore within the megasporangium, which is enclosed within the ovule. Study the prepared slide of a cross-section through the **ovary** of Lilium. Note that the ovary consists of three carpels in each of which are two ovules. Each ovule consists of a stalk -- the **funiculus** -- by which it is attached to the ovary wall; that portion of the ovary wall to which an ovule is attached is the **placenta**. The **nucellus** sporangium wall) is enclosed in maternal sporophytic tissue, the **integuments**. At one end of the ovule, the integuments are incomplete, leaving an opening -- the **micropyle** to the nucellus. In early stages of development the megasporangium contains a single **megaspore mother cell** that ultimately undergoes meiosis, giving rise to four **megaspores**. Three of the four megaspores within a megasporangium disintegrate. The remaining megaspore undergoes mitosis > 2 haploid nuclei; an additional mitosis > 4 haploid nuclei; a final mitotic division > 8 nuclei. These eight nuclei contained within a megasporangium within the ovule constitute the mature megagametophyte of the angiosperm, often referred to as an embryo sac. The eight nuclei become arranged within the megasporangium such that three -- the **antipodal nuclei** -- appear at the end of the ovule opposite the micropyle, i.e., the **chalazal** end. Two nuclei -- the **polar nuclei** -- form the **central cell** near the center of the embryo sac. Three nuclei become organized at the **micropylar end** of the ovule; two of these represent the **synergid nuclei**, the remaining nucleus is the **egg nucleus**. To complete your study of megasporogenesis and development of the female gametophyte, study the prepared slides and models on angiosperms. The demonstration shows various stages in egg development.

Sketch an ovule with a mature female gametophyte and label: **placenta, funiculus, integument, nucellus, micropyle, antipodals, polar nuclei** (central cell), **synergids**, and **egg cell**.

Figure 9.2. Pistil cross section and development of the ovule.

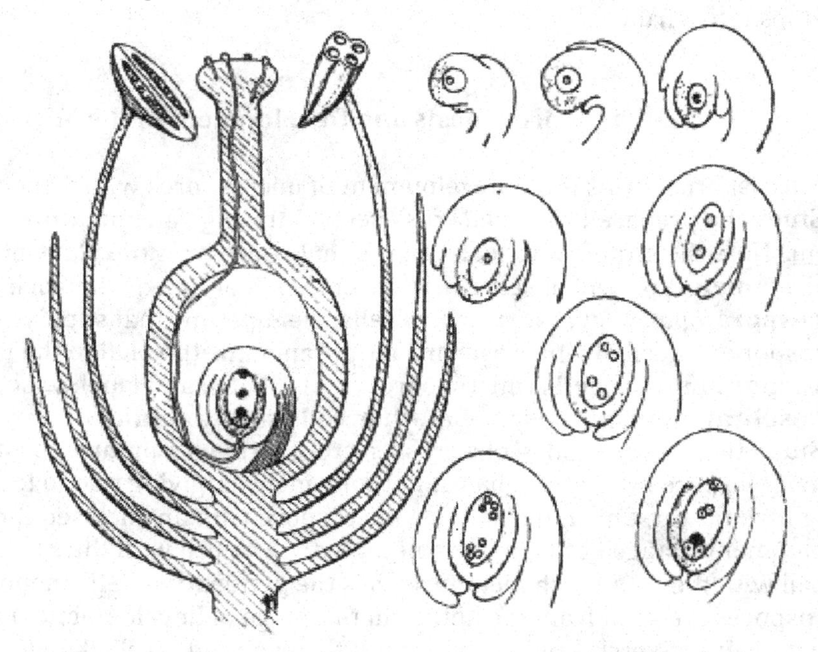

Pollination and Fertilization

Upon rupture of the microsporangium walls of the anthers, the pollen grains are transferred to the stigmas in a variety of ways: wind, insects, birds, mammals, etc. The process of transferral is referred to as **pollination**. When a pollen grain lands on a compatible stigma, a **pollen tube** begins to grow out of the grain. If the **generative nucleus** of the microgametophyte has not previously divided to form two **sperm** (flagellated), it does so now. The mature male gametophyte consists of a pollen grain with its tube nucleus and two sperm.

The pollen tube grows through the stigma and style and, eventually, penetrates the nucellus, usually at the micropylar end of the ovule. The tube nucleus disintegrates, and the two sperm nuclei enter the **embryo sac**. Upon entering the embryo sac, one sperm fuses with the two **polar nuclei** (or central cell), forming the primary **endosperm nucleus**. The other sperm nucleus fuses with the **egg nucleus**, forming the **zygote**. This process, involving the fusion of both sperm produced by the microgametophyte with elements of the megagametophyte is commonly referred to as **double fertilization** and, except for its occurrence in two gymnospermous genera, is unique to angiosperms. After double fertilization, the ovule develops into a **seed**.

Demonstration:
Study the prepared slide of a "germinated" pollen grain. Note the remnants of the microspore and the pollen tube within which you can see "leading" **tube nucleus** and the two "trailing" **sperm nuclei**. Sketch and label: **spore wall, tube nucleus,** and **sperm nuclei**.

The Mature Embryo Sporophyte and Seed

After double fertilization, the primary endosperm nucleus begins to divide and gives rise to a nutritive tissue, the **endosperm**, which surrounds the **embryo sporophyte** has concomitantly developed from divisions of the **zygote**.

The mature **embryo sporophyte** is composed of a stem-like axis bearing: (1) one [in representatives of the monocotyledonous angiosperms] or two [in representatives of the dicotyledonous angiosperms] **cotyledons** ("seed leaves"); (2) an **epicotyl** [above the point of attachment of the cotyledon(s)] with the **shoot apical meristem** at its apex and (3) a **hypocoty-root axis** [below the point of attachment of the cotyledon(s)] with the **apical meristem** (radicle) at its apex.

The post-fertilization ovule -- the **seed** -- of the angiosperm thus consists of the **seed coat**, derived from the integuments (plus, in some cases, remnants of the nucellus), the **embryo sporophyte**, derived from the zygote and the **endosperm** derived from the primary endosperm nucleus. The amount of endosperm in angiosperm seeds varies; in some cases, the endosperm is "used up" very quickly by the developing embryo.

Study the prepared slide of a longitudinal section of a seed of Capsella. Label the following figure: **seed coat, cotyledons, shoot apical meristem, hypocotyls-root axis,** and **root apical meristem**.

Figure 9.3. Cross section of a seed.

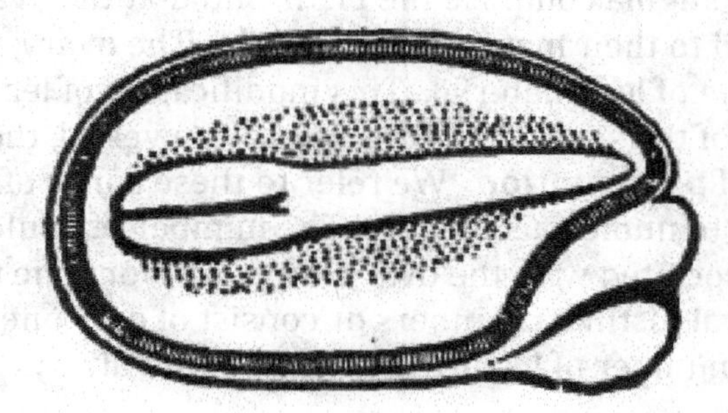

Obtain a garden bean that has been previously soaked in water. Note the following superficial features, sketch and label the **seed coat** and **micropylar scar**.

Carefully remove the seed coat of the garden bean seed, dissect and examine with a dissecting microscope. Sketch and label; **seed coat, epicotyl** (with shoot apical meristem), **plumule** ("foliage leaves"), **cotyledons**, and **hypocotyl-root axis** (with root apical meristem).

EXPERIMENT: Will Seeds Germinate well without a Good Supply of Air?
Place some soaked seeds on damp blotting paper in the bottom of a bottle, using seeds enough to fill it three-quarters full, and close tightly with a rubber stopper. Place a few other seeds of the same kind in a second bottle; cover loosely. Place the bottles side by side, so that they will have the same conditions of light and heat. Watch for results, and tabulate as in previous experiments. Most seeds will not germinate under water, but those of the sunflower will do so, and therefore Exp. Ill may be varied in the following manner: Remove the shells carefully from a considerable number of sunflower seeds. Try to germinate one lot of these in water which has been boiled in a flask to remove the air, and then cooled in the same flask. Over the water, with the seeds in it, a layer of cottonseed oil about a half inch deep is poured, to keep the water from contact with air. In this bottle then there will be only seeds and air-free water. Try to germinate another lot of seeds in a bottle half filled with ordinary water, also covered with cotton-seed oil.

Results?

The Mature Fruit

INTRODUCTION: The goal of this laboratory is to turn your senses and intellect on to the phenomenal structures commonly known as **fruits**. Changes in the ovary wall occur simultaneously with the maturation of the seed, giving rise to a **FRUIT**. In flowers, we observed that the ovary, style, and stigma compose the pistil, and that the ovary is a protective vessel in which ovules are nourished to their mature form--**seeds**. The ovary, with time, evolved through the constant modification of leaf material. This modification of leaf material provides protection and facilitates dispersal of the seeds. Within this ovarian vessel, the ovules remain attached to parent tissue along zones of **placentation**. We refer to these zones of placentation as **carpels**. Ovaries can be composed of one to numerous carpels. The number of ovules associated with each carpel, and thus the number associated with the ovary, can vary from one to many. Also, ovaries can be separated into several distinct chambers or consist of only one chamber. These chambers are called **locules**. The number of locules is often (but not always) equal to the number of carpels.

Note that a number of common foods in human diets are erroneously classified as "vegetables" when they are actually fruits! Vegetables are only plant parts that are vegetative (non-reproductive) (example carrot, lettuce, celery, etc).

Basic parts of a fruit of a tomato and orange:

1. **Pericarp** - the fruit wall (composed of #2, #3, #4).

2. **Ectocarp** or **Exocarp** - the outermost layer of the pericarp.
3. **Mesocarp** - the middle layer of the pericarp.
4. **Endocarp** - the inner layer of the pericarp.
5. **Placenta** - a region of attachment of seeds on the fruit wall.
6. **Funiculus** - the stalk attaching the seed to the placenta.
7. **Seed** - a matured ovule.

Figure 9.4. Cross section of a tomato and an orange.

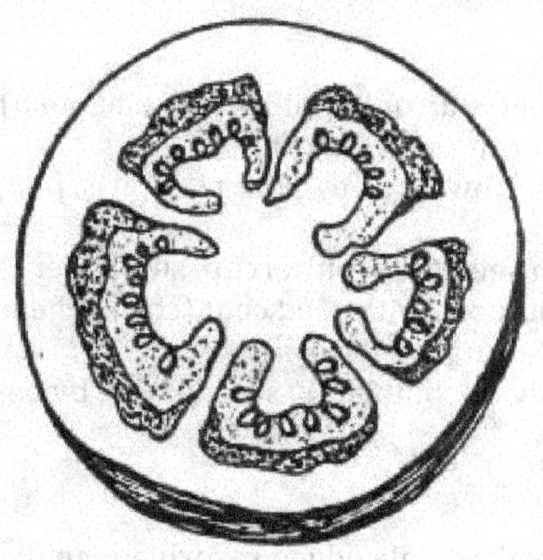

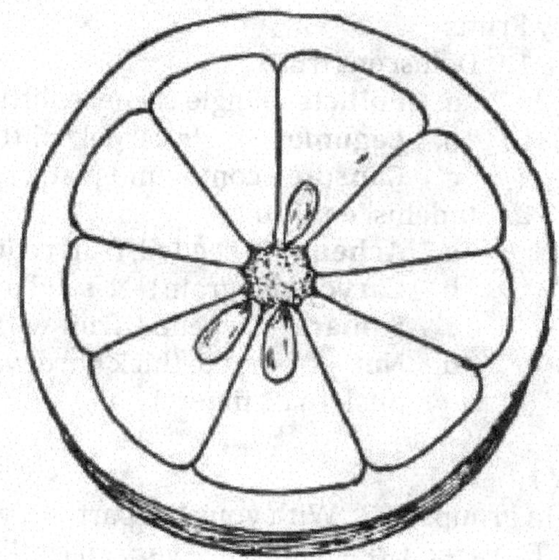

Fruit structure has been greatly influenced by the mode of seed dispersal. Seeds can be dispersed by animals, wind, water, etc. Fruits are a major portion of the diet of animals. Animal dispersers range from insects to birds; mammals to fish. Modifications in the shape, structure, and often color of the protective ovary directly correspond to the ways in which seeds are dispersed. Fruit-eating birds commonly disperse bright red, fleshy berries. Winged fruits, such as those found on maple trees, have obviously come about through modifications that facilitate wind dispersal. How might nuts, such as acorns, be dispersed? What animals have you observed eating and burying acorns in the autumn?

By the end of this laboratory exercise you should be able to recognize the various fruit types that are present in nature, and understand the similarities and differences between these fruit types. You should become familiar with the terminology used to describe fruits, such as carpel, ovule, zones of placentation, etc. You should be able to count the number of carpels present in a given fruit and to recognize the arrangement of these carpels within the ovary. Lastly, and most importantly, you should be able to use the Fruit Key to identify unknown fruit types. You should leave the laboratory with a deeper appreciation for the fruits of the flowering plant world and a stomach full of these incredibly delicious phenomena!

Three Fruit Classification Schemes

ORIGINS

A) **Simple** fruit - formed from a single pistil (lily, apple, cucumber)

B) **Aggregate** fruit - formed from a cluster of separate pistils borne in a single flower (raspberry)

C) **Multiple** fruit - formed from the pistils of several to many flowers consolidated with other floral or inflorescence parts (pineapple, fig)

COMPOSITION

A) **True** fruit - composed of only the ripened ovary, with its contained seeds (lily)

B) **Accessory** fruit - composed of the ripened ovary with other additional parts, such as receptacle, bracts, portions of perianth, etc. (apple, cucumber, fig)

DESCRIPTIONS

A) Fleshy Fruits

1. **Drupe** - usually 1-seeded, fruit coat with fleshy outer and inner stony layers (peach, plum, olive, raspberry, almond)
2. **Berry** - few to many seeded, fruit coat soft and fleshy throughout (grape, banana)
   a. **Hesperidium** - berry with tough rind (orange, grapefruit)
   b. **Pepo** - thick-skinned berry, accessory (squash, cucumber)
3. **Pome** - fleshy accessory fruit with cartilaginous core (apple, pear)

B) Dry Fruits

1. Dehiscent fruits
   a. **Follicle** - single carpel splitting along one side only (milkweed, magnolia)
   b. **Legume** - single carpel splitting along both sides (bean)
   c. **Capsule** - compound pistil, splitting lengthwise or by pores (lily, iris, poppy)
2. Indehiscent Fruits
   a. **Achene** - 1-seeded, fruit coat free from seed coat (buttercup, sunflower)
   b. **Caryopsis** (grain) - 1-seeded, fruit coat fused with seed coat (corn, wheat)
   c. **Samara** - 1-seeded, fruit with wing-like outgrowth (ash)
   d. **Nut** - 1-seeded, thick hard wall, partially or completely surrounded by cup or husk (oak, filbert)

ACTIVITY I

Work in groups of 2. With your lab partner, you will need a single edged razorblade to dissect the fruits. Be careful, the blades are very sharp!!! Before you start dissecting and possibly consuming your lab material, you will need to be able use the dichotomous key (on page 81) to identify each fruit. After you have identified all of the known and unknown fruits, try to observe all of the following structures (if present): **pericarp, exocarp, mesocarp, endocarp, placenta, funiculus, seed, samara, locules, and zones of placentation.**

Table 9.2. Examine the variety angiosperm fruits on display in the laboratory

Common Fruit Name	Fruit Type
Apple	
Tomato	
Grass seed	
Sunflower Seed	
Maple Seed	
Green Bean	
Acorn	
Cucumber	
Orange	
Peanut	
Banana	
Unknown Fruit #1	
Unknown Fruit #2	
Unknown Fruit #3	
Unknown Fruit #4	
Unknown Fruit #5	

FILL IN THE APPROPRIATE SECTIONS IN THE TABLE ON PAGE 78

LABORATORY 10.
ADULT PLANT CELLS AND TISSUES OF THE HERBACEOUS STEM

Among angiosperms, considerable variation exists in the structure of stems, leaves and roots, but this variation is largely quantitative rather than qualitative. All three organs are comprised of the same tissue systems, tissues and types of cells. The tissue systems -- **dermal, ground**, and **vascular** -- are initiated during embryological development as a function of cell differentiation and are continuous with one another in all tissue systems of the same individual. These tissue systems are initiated during development of the embryo sporophyte where they are represented in the shoot and root apical meristems by three distinct cellular regions ("germ layers") -- sometimes called primary meristems): the **protoderm, ground meristem**, and **procambium**, respectively. As new cells are produced by the apical meristems, cells differentiated from the protoderm become **epidermis,** i.e., the **primary dermal tissue**; those differentiated from the ground meristem become **cortex** and, when present, i.e., the **ground tissues**; those differentiated from the procambium become the **phloem**; i.e., the **vascular tissues**. By definition, all cells and tissues derived from divisions of an apical meristem -- shoot or root -- are **primary tissues.**

Cells and Primary Tissues of the Herbaceous Stem

Obtain a prepared slide of a longitudinal section of the stem (shoot) apex of Coleus, a herbaceous dicot. Identify and label on the following figures: **shoot apical meristem, leaf primordium**, and **axillary (lateral) bud primordium**. Recall that the apical meristem has three primary meristems ("germ layers") -- **protoderm, ground meristem**, and **procambium** -- that are responsible, respectively, for giving rise to new cells differentiating as epidermis, **dermal tissue** (epidermis), **cortex, ground tissue** (pith) and **primary phloem, primary xylem**, and **vascular tissues.**

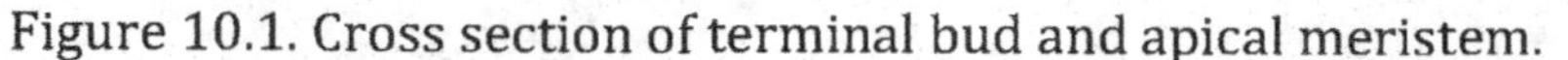

Figure 10.1. Cross section of terminal bud and apical meristem.

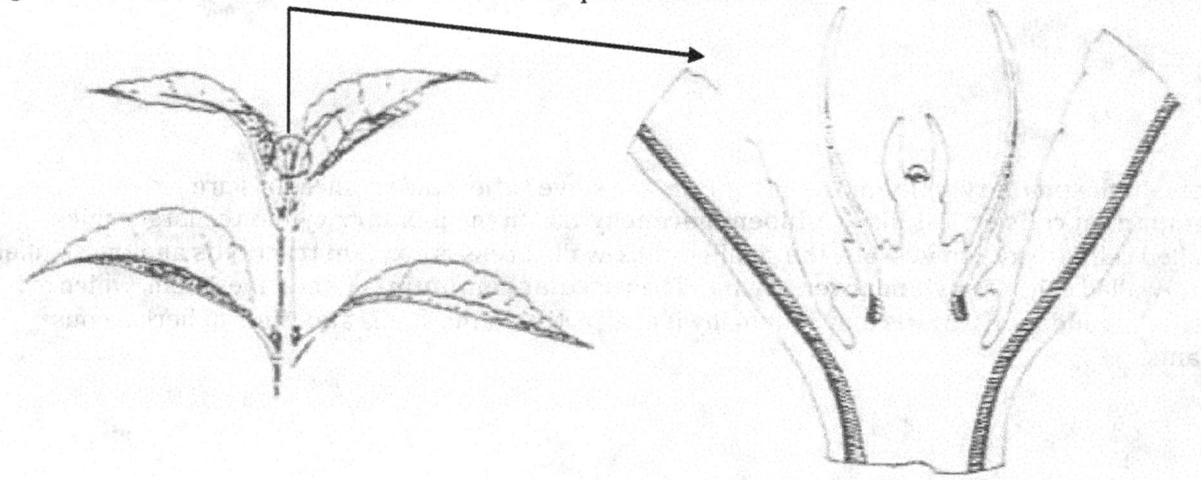

Obtain a prepared slide of a cross-section of the stem of Helianthus, an herbaceous dicot. Identify and label on the following figures: **epidermis, cortex, pith ray, vascular bundle cap, sclerenchyma/primary phloem fibers, primary phloem, primary xylem, cuticle, epidermis, collenchyma, parenchyma, cortex, vascular bundle cap, sclerenchyma, primary phloem fibers, primary phloem, vascular cambium, primary xylem**, and **pith ray.**

Figure 10.2. Cross section of a herbaceous dicot stem.

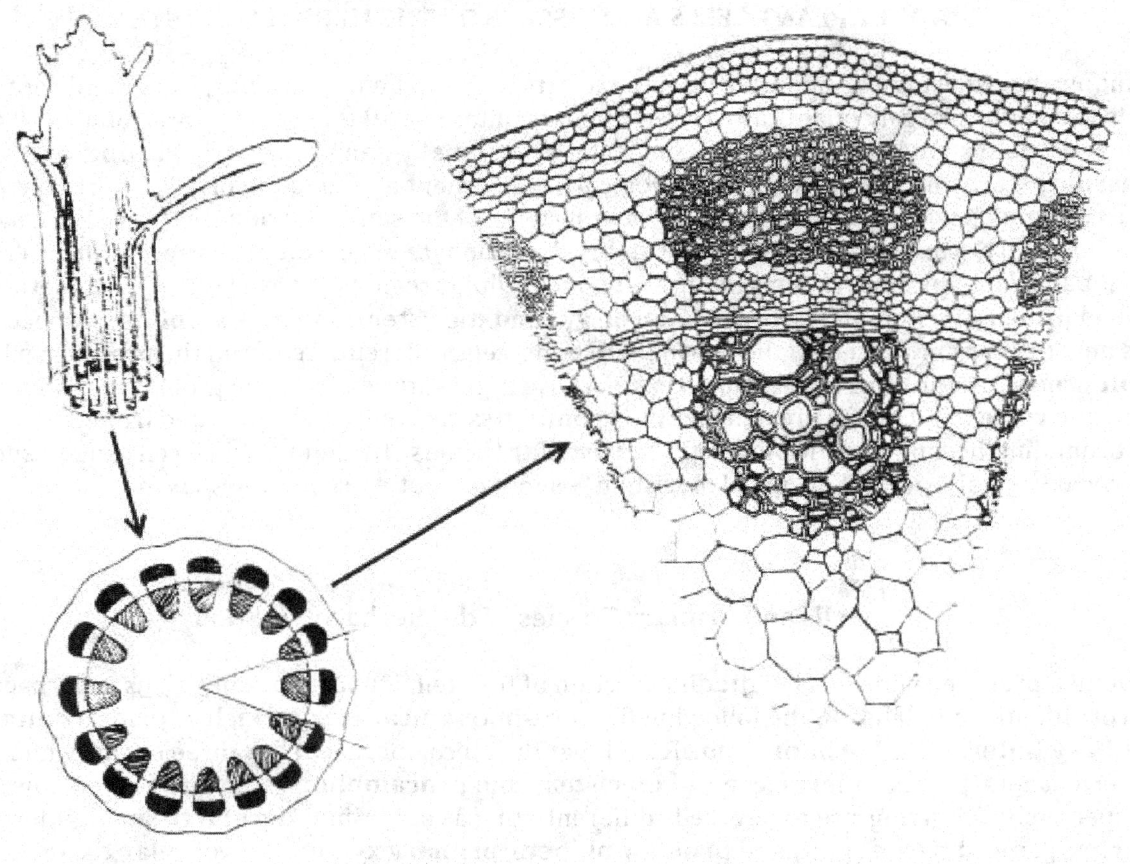

Note: In the **primary phloem** the larger cells are **sieve tubes**; the smaller cells are either **companion cells** or, less likely, **phloem parenchyma**. In the **primary xylem** the larger thick-walled cells are **xylem vessels**; the smaller thick-walled cells are **xylem tracheids** and the smaller thin-walled cells are **xylem parenchyma**. The **vascular cambium**, a lateral meristem, which gives rise to secondary tissues, is only minimally if at all active in the stems and roots of herbaceous plants.

Obtain a prepared slide of a cross-section of the stem of <u>Zea mays</u>, an herbaceous monocot.

Identify and label on the following figures: **epidermis, ground tissue, parenchyma, vascular bundle sheath, sclerenchyma, primary phloem, primary xylem**, and **air space**.
Note: In the **primary phloem** the larger cells are **sieve tubes**; the smaller cells are either **companion cells** or, less likely, **phloem parenchyma**. In the **primary xylem** the larger thick-

walled cells are **xylem vessels**; the smaller thick-walled cells are **xylem tracheids** the smaller thin-walled cells are **xylem parenchyma**. There is no vascular cambium.

Figure 10.3. Cross section of a monocot stem.

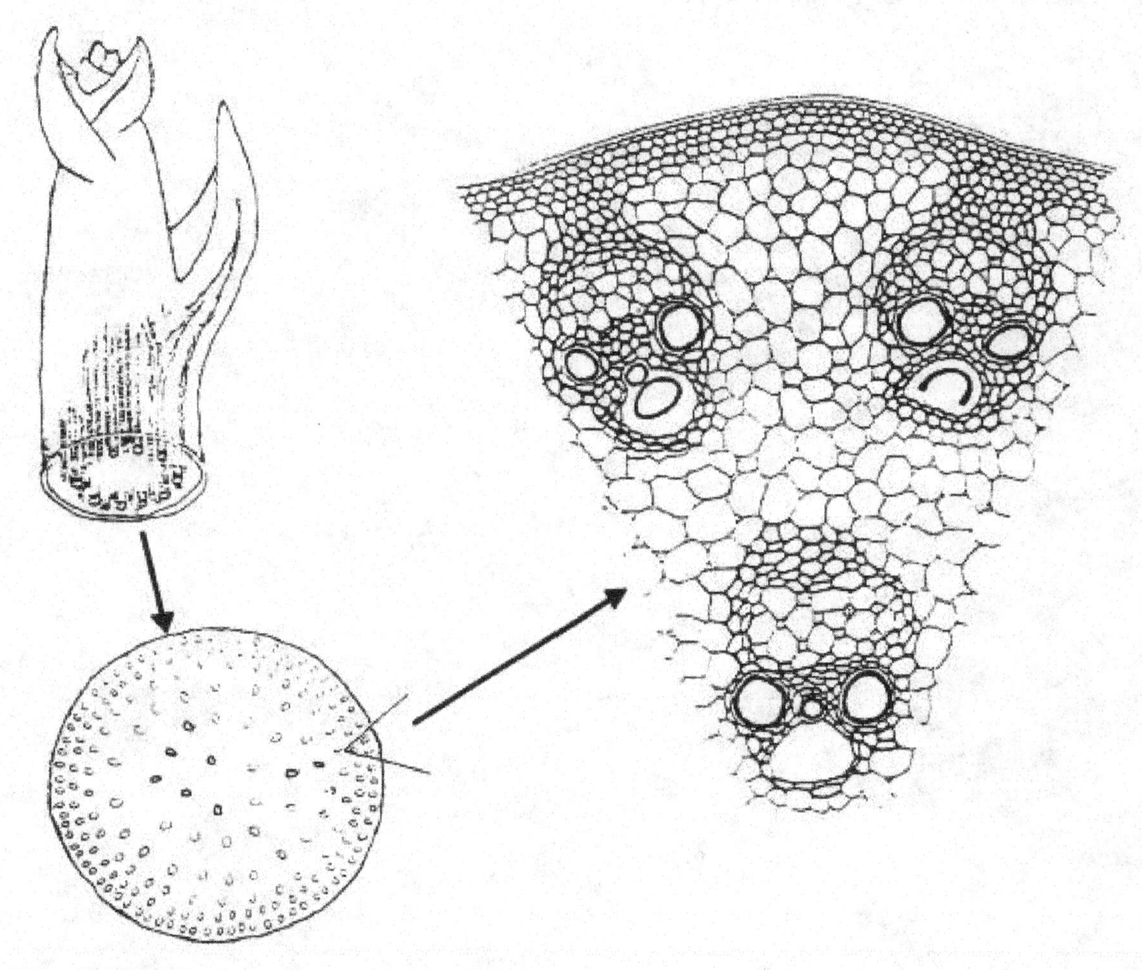

Figure 10.4. Monocot stem vascular bundle.

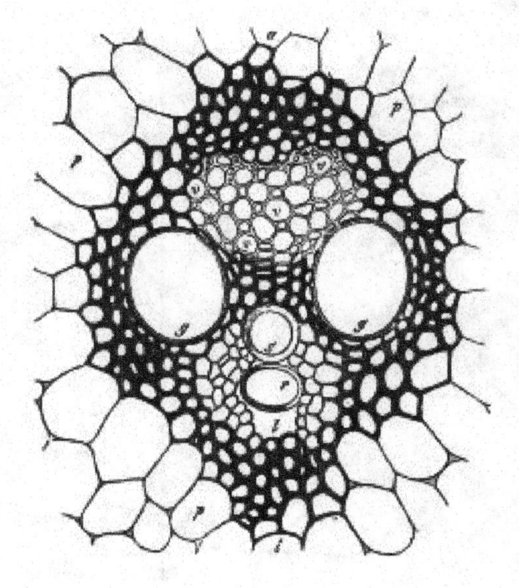

LABORATORY 11.
WOODY STEMS

During the last laboratory meeting you completed a study of the anatomy of herbaceous stems. As you will recall, the growth of herbaceous plants is largely primary in nature, i.e., all, or nearly all, of the growth is due to the activity of apical meristems, which by definition, produce primary tissues. All plants exhibited primary growth, both those that are herbaceous and those that are woody. However, in woody plants the "soft" tissues of the stem and root produced during primary growth are later replaced with other tissues that make the stem and root "hard". These tissues are the result of secondary growth that is due to the activity of one, or more, **lateral meristems**.

Thus, a woody plant is one that develops secondary tissues, giving rise to a permanent structure that yearly increases its leaf canopy through continued primary growth.

A woody stem has three basic parts composed as shown below:

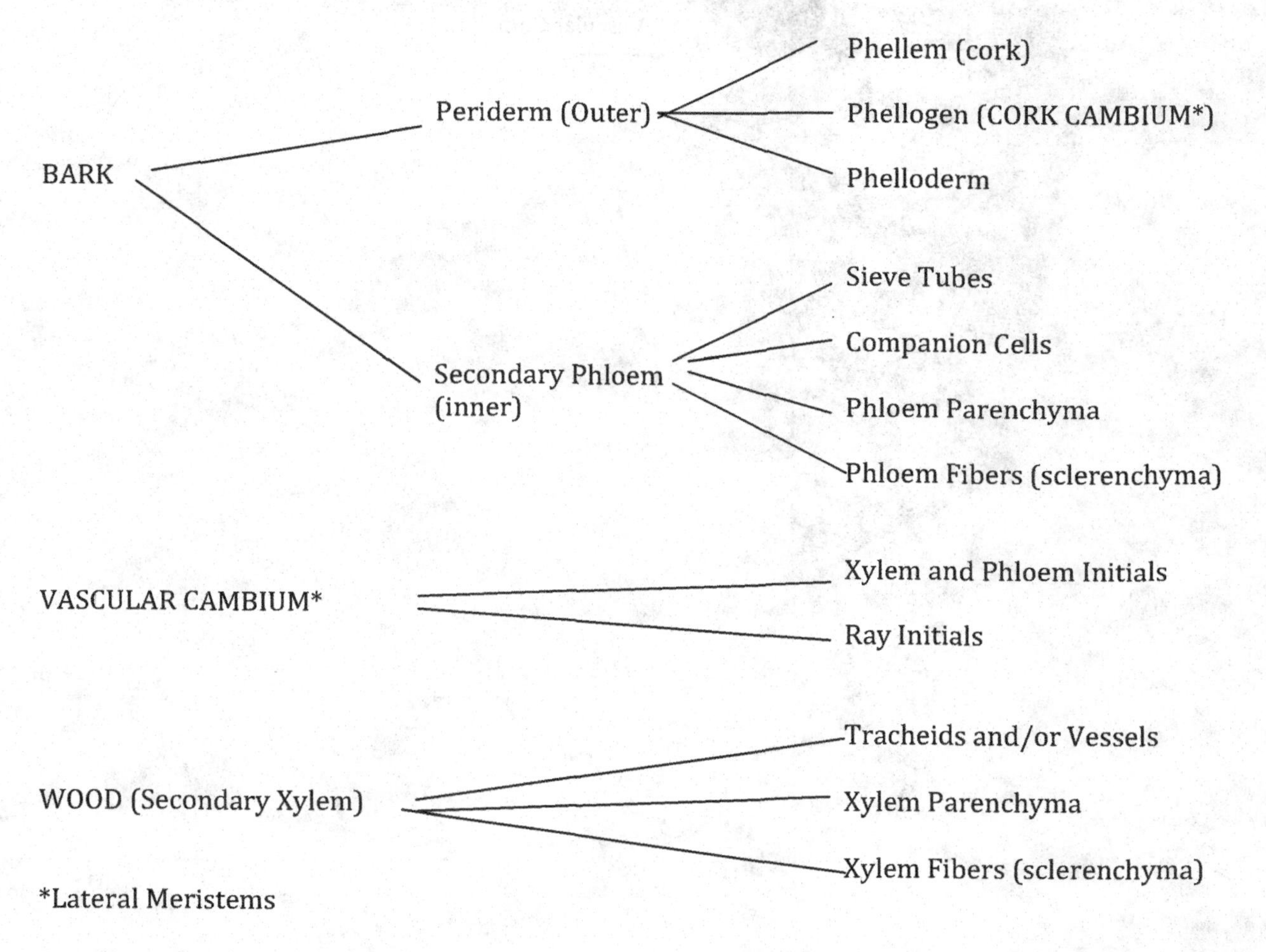

*Lateral Meristems

Demostration:

Obtain a prepared slide of macerated angiosperm (secondary xylem), probably of oak, Quercus. Identify and sketch the four types of cells comprising the secondary xylem, keeping in mind that these cells are secondary in origin, having been produced as a result of activity of the **vascular cambium**. Remember, however, that these same types of cells comprise the primary xylem though; in that case, the cells are derived from the **shoot apical meristem**.

Obtain a prepared slide of a cross-section of the stem of the woody dicots, Tilia and Quercus. Note that the slide has four tissue sections representing stems of differing age. As you identify and study the following cells/tissues – **Epidermis, Cortex, Primary Phloem, Secondary Phloem, Phloem**

Rays, vascular cambium, Primary xylem, Secondary xylem, Xylem rays, pith, Annual rings (growth rings), summerwood, springwood, sapwood, heartwood.

Figure 11.1. Cross section of a Woody dicot stem.

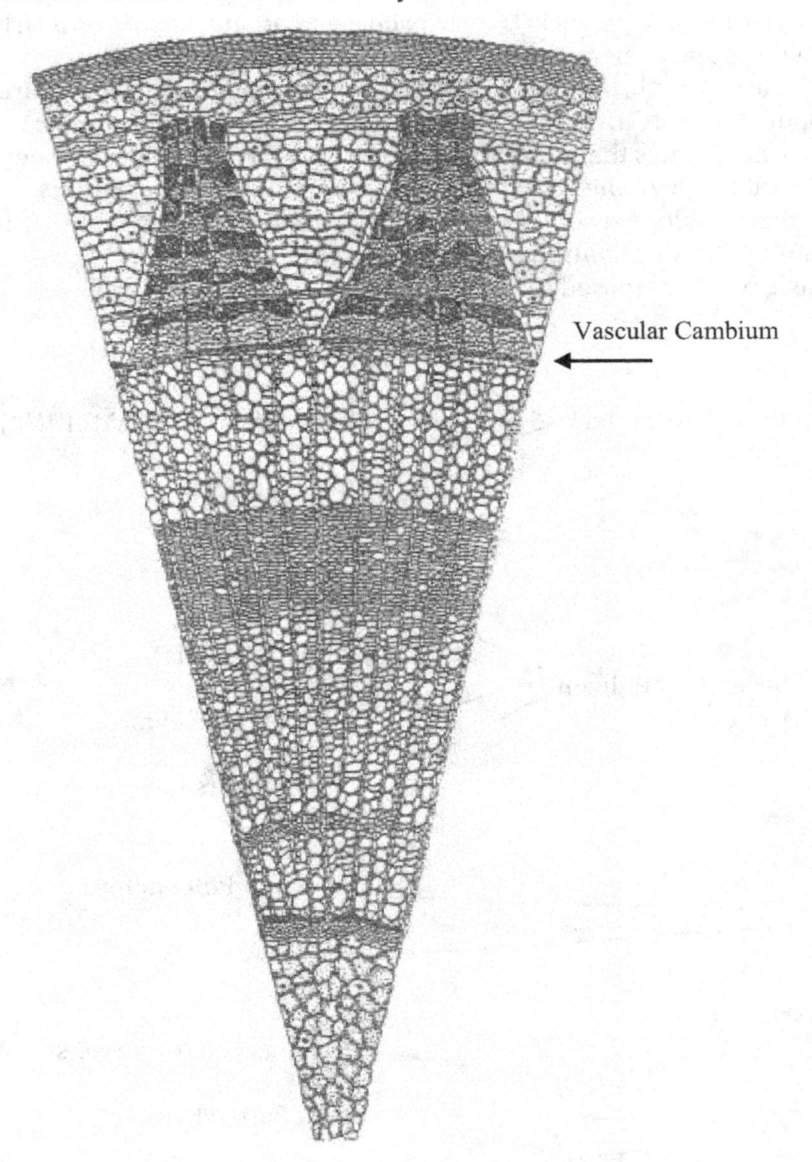

LENTICEL

Note on the surface of the twig specimen the presence of small, rounded (pinhead-like) or linear (slit-like) markings. These are the **lenticels, which** serve as avenues for gas exchange by the woody stem just as stomata provide for gas exchange in leaves and young stems. Remember that the **epidermis**, a primary tissue, is replaced by **phellem** and **phelloderm** as the twigs of woody plants mature.

Figure 11.1. Cross section of a lenticel in bark.

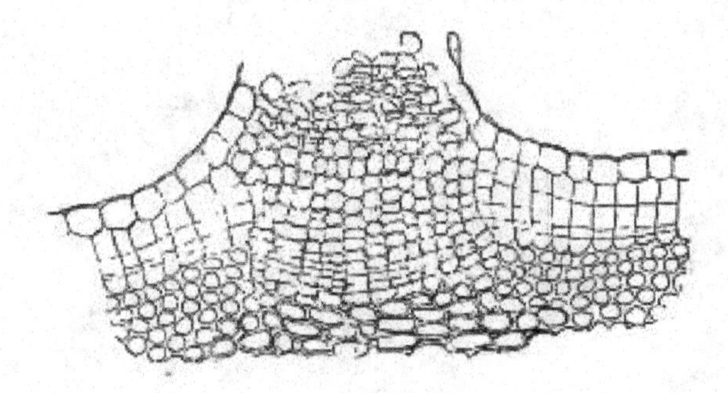

External Features of Woody Stems

Obtain a woody twig; twigs are the youngest parts of the shoot system of woody plants. Identify the following and label the drawing above, appropriately: **Terminal bud, apical meristem, bud scales, Leaf scars, Vascular Bundle scars, node region, internode region, lateral buds, terminal bud scale scar, lenticels, epidermis or cork cells.**

Figure 11.2. Winter woody twig.

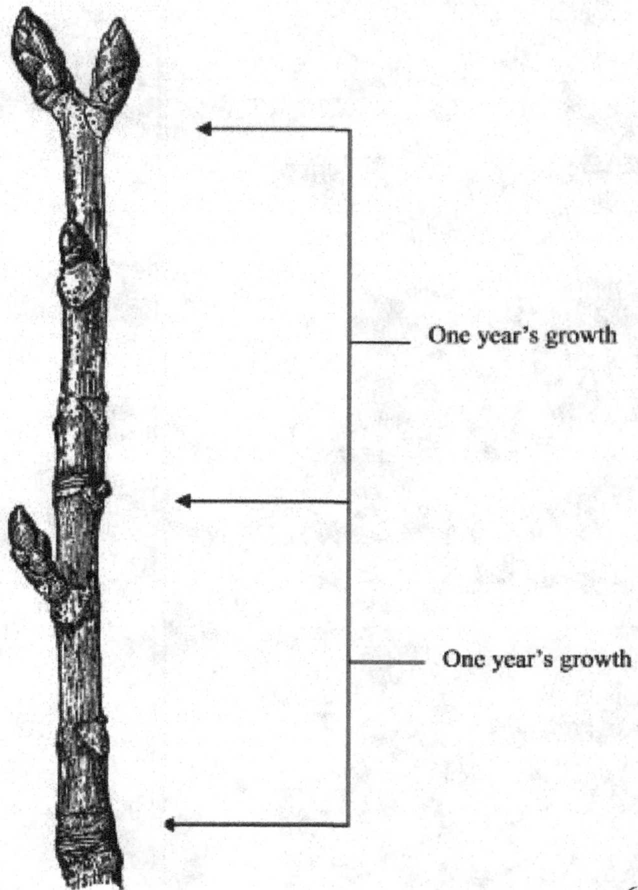

LABORATORY 12.
LEAVES and ROOTS

Leaves - The Major Organs of Photosynthesis

Even though vascular plants are well adapted to life on land, photosynthetic reactions require an aqueous medium, a "holdover" from the aquatic ancestral charophycean algae. Obviously, a terrestrial plant is not surrounded by water. The carbon dioxide necessary for photosynthesis is not found as a free gas in the cells but is dissolved in water. The leaf of a plant is basically a layer of photosynthetically active cells "sandwiched" between two protective layers -- the upper and lower epidermis. As you know, the epidermis has guard cells and stomata, which allow for gas exchange with the atmosphere.

Among angiosperms, largely in response to the specific environment to which the plant is adapted -- hydric, mesic or xeric -- there is considerable variation in leaf morphology and anatomy. You will study the anatomy of a leaf which is rather typical of those produced by plants occurring in mesophytic habits in climatically temperate zones.

Review the study you made of the stem apex (Laboratory 10) in respect to the origin of leaves!

Leaf Anatomy

Obtain a prepared slide of a cross-section of a leaf of a mesophytic dicot, probably (privet). Identify and label the following figure: **upper epidermis, cuticle, cutin, lower epidermis, guard cell, palisade parenchyma, spongy parenchyma,** and **intercellular space**. Study the mid-vein in detail, identify and label: **vascular bundle sheath, phloem,** and **xylem.**

Figure 12.1. Cross section of a dicot leaf.

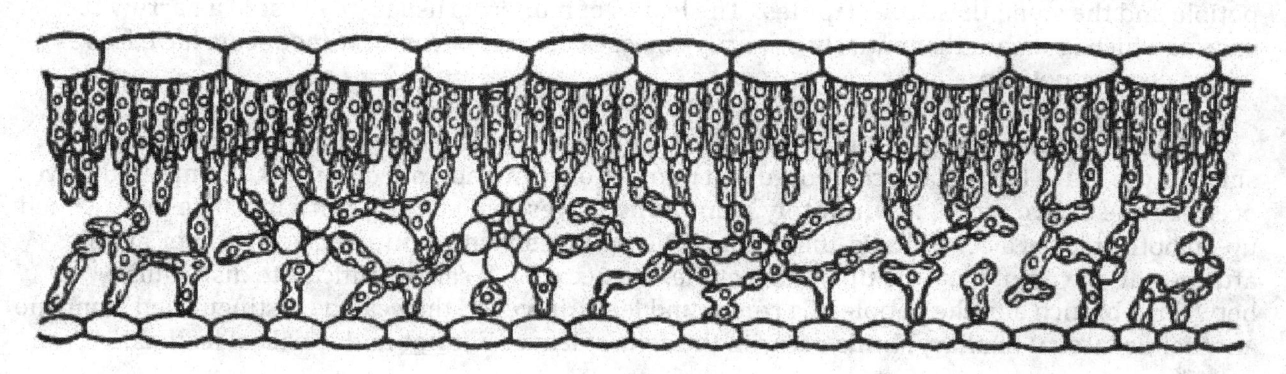

Figure 12.2. (A) Epidermal cells with stomata pores (B) Cross section of a monocot leaf.

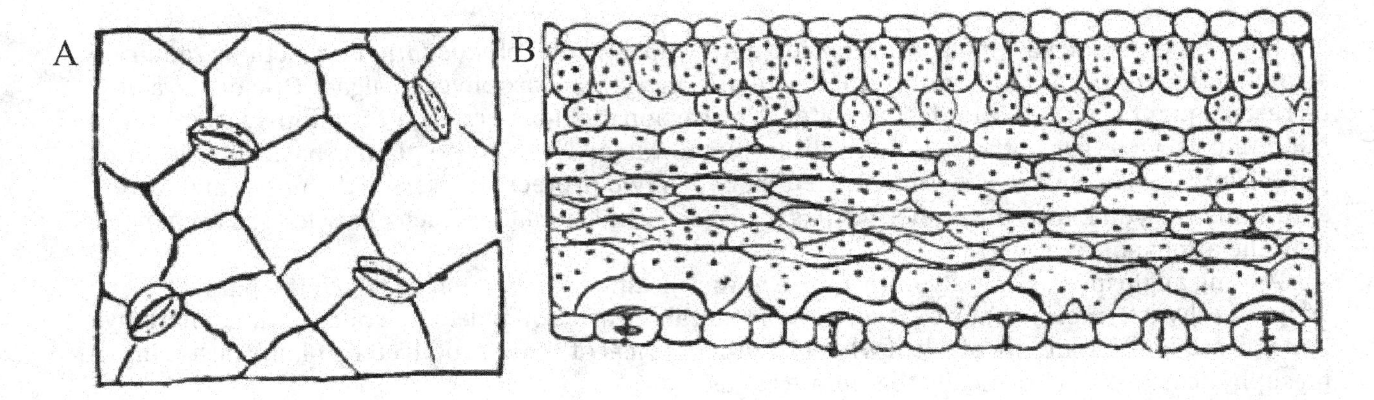

Leaf External Features

Study the material available in the laboratory in respect to the following external features of angiosperm leaves. Make sketches, as you deem necessary. The leaves of dicots consist of a flattened laminar portion, the **blade** (lamina) and a **stalk**, the **petiole**, by which the blade is attached to the stem. Some plants have one or two small, leaf-like appendages at the junction of the petiole and the stem; these are **stipules**. The leaves of monocots usually consist of a narrow the base of which is sheath that entirely or partly encloses the stem. Very few monocots have leaves with a distinct **petiole**.

Leaves are either **simple** or **compound**. The former has a blade consisting of one expanded surface; the latter has a blade composed of a number of segments called **leaflets**. Compound leaves occur as one of two types. In **pinnately compound** leaves, leaflets, occur in a linear sequence lined up on both sides of a central axis, the **rachis**. The **palmately compound** leaves, leaflets, are all attached at one point near the tip of the petiole. On occasion it can be difficult to distinguish between a branch and the petiole of a compound leaf. However, they can be distinguished from one another as **lateral** (axillary) **buds** occur in the axils of leaves but **not** in the axils of leaflets!

Leaves are attached to the stem in different patterns. In **alternate** arrangement leaf occurs at each node. In **opposite** arrangement two leaves occur at each node. In **whorled** arrangement three or more leaves occur at each node.

Venation, the pattern of veins in leaves, though variable among different species, falls into two general categories: **net venation** and **parallel venation**. The former is generally characteristic of dicots; the latter is generally characteristic of monocots. In **net venation** there is one, or more, large main veins from which arise smaller veins that branch and anastomose, forming a net-like system. There are two patterns of net venation: **pinnate net venation** and **palmate net venation**, both of which have a network of smaller veins between the main vein(s). In the former there is one main vein extending from the base of the leaf blade to its apex. In the latter there are several large main veins that radiate from a common point of origin at the base of the leaf blade. In **parallel venation** numerous main veins are parallel to one another and extend the length of the leaf blade without a network of smaller veins in between.

Roots -- Organs of Anchorage, Absorption and Translocation

Study the material available in the laboratory in respect to the following major types of root systems. Make sketches, as you deem necessary.

Among angiosperms there occur three major types of root systems:

TAP ROOT

There is one predominant root that bears smaller lateral (branch) roots. You will recall that it originated as the primary root from the apical meristem of the hypocotyls/root axis of the embryo.

FIBROUS ROOT

There are numerous equal-sized roots originating from the apical meristem of the hypocotyl/root axis of the embryo.

ADVENTITIOUS ROOT

Though similar in general appearance to fibrous roots, adventitious roots originate from stems and/or leaves, from the apical meristem of the hypocotyl/root axis of the embryo; i.e., they are embryonic in origin but arise from mature tissues. The capacity of stems to generate adventitious roots is the basis for the propagation of ornamental plants by stem cuttings.

Though the older portions of roots are absorptive to some extent, water absorption is largely a function of younger portions of the root system and by **root hairs** located near the root apex. The young, absorptive apices are known as **feeder roots**. In most trees, the feeder roots lie within 15-20 cm of the soil surface, a depth at which oxygen, water and nutrients are readily available.

Demonstration:

Examine the root of a young seedling of radish (Raphanus sativus) on demonstration under a dissecting microscope. Note the white, "fuzzy" appearance of the primary root, which is due to the presence of numerous **root hairs**. Each root hair is a tube-like outgrowth of an individual epidermal cell. They significantly increase the absorptive surface area of the root.

Figure 12.3. Epidermis layer with root hairs.

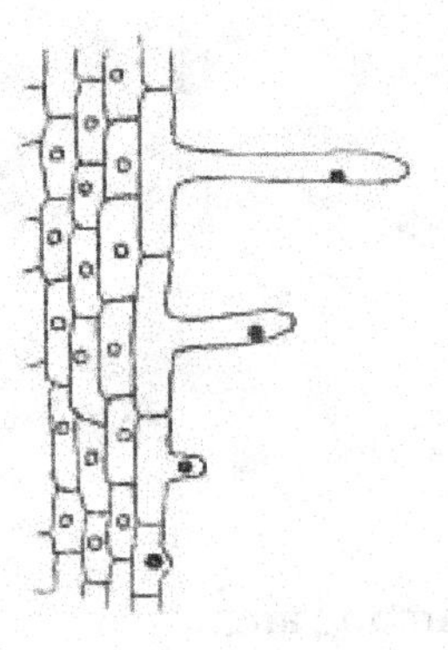

Cells and Primary Tissues of the Root

Obtain a prepared slide of a longitudinal section of the root tip of Onion (Allium cepa). Identify and label (**BELOW**) those structures appearing below in **bold face**.

In previous study, you saw that the enveloping cotyledons protected the apical meristem of the shoot apex of a germinating garden bean seed. You will also have noted in your study of the stem apex of Coleus that later the leaf primordial "overlap" the apical meristem. The apical meristem of the root apex, an equally delicate tissue, also needs protection from the abrasive action of soil particles as it grows. A mantle of cells on the root apical meristem, the **root cap**, provides this protection.

Immediately behind the **root cap** is the **root apical meristem** with its three primary meristems, the **protoderm, ground meristem**, and **procambium**. REVIEW the functions of these three primary meristems as studied.

Immediately behind the **root apical meristem** is the **region of elongation**. It is in this region that new cells produced by the apical meristem undergo elongation due chiefly to internal pressures that build up by an increase of water in the large central vacuoles of each cell. Cell elongation increases root length is the force "pushing" the tip of the root forward (downward).

Immediately behind the region of elongation is the **region of maturation**. Note: This region may not be represented in the section on your slide. Externally, this region is marked by the appearance of **root hairs**. It is in this region that cells become differentiated into the primary tissues of the root.

EXPERIMENT

Root-Hairs. Barley, oats, wheat, red clover, or buckwheat seeds soaked and then sprouted on moist blotting paper afford convenient' material for studying root-hairs. The seeds may be kept covered with a watch-, glass or a clock-glass while sprouting. After they have begun to germinate well, care must be taken not to have them kept in too moist an atmosphere, or very few root-hairs will be formed. Examine with the magnifying glass those parts of the root which have these appendages. Try to find out whether all the portions of the root are equally covered with hairs and, if not, where they are most abundant. The root-hairs in plants growing under ordinary conditions are surrounded by the moist soil and wrap themselves around microscopical particles of earth. Thus they are able rapidly to absorb through their thin walls the soil-water, with whatever mineral substances it has dissolved in it.

Because of the regular occurrence of cell division (mitosis/cytokinesis) in the apical meristem of the root tip of *Allium cepa*, this tissue is often used to study the stages in mitotic nuclear division. REVIEW the stages of mitosis and identify those stages in your study of the apical meristem of the onion root tip.

Obtain a prepared slide of a cross-section of the root of the dicot, Ranunculus (buttercup). Identify and label the following cells/tissues on the figure above: Epidermis, Cortex, **Parenchyma, Starch grains,** Vascular Cylinder (stele), **Endodermis, Casparian strips, suberin layer,** Pericycle, **Primary xylem, Primary phloem, Vascular cambium.**

Figure 12.4. Cross section of a dicot root and vascular cylinder.

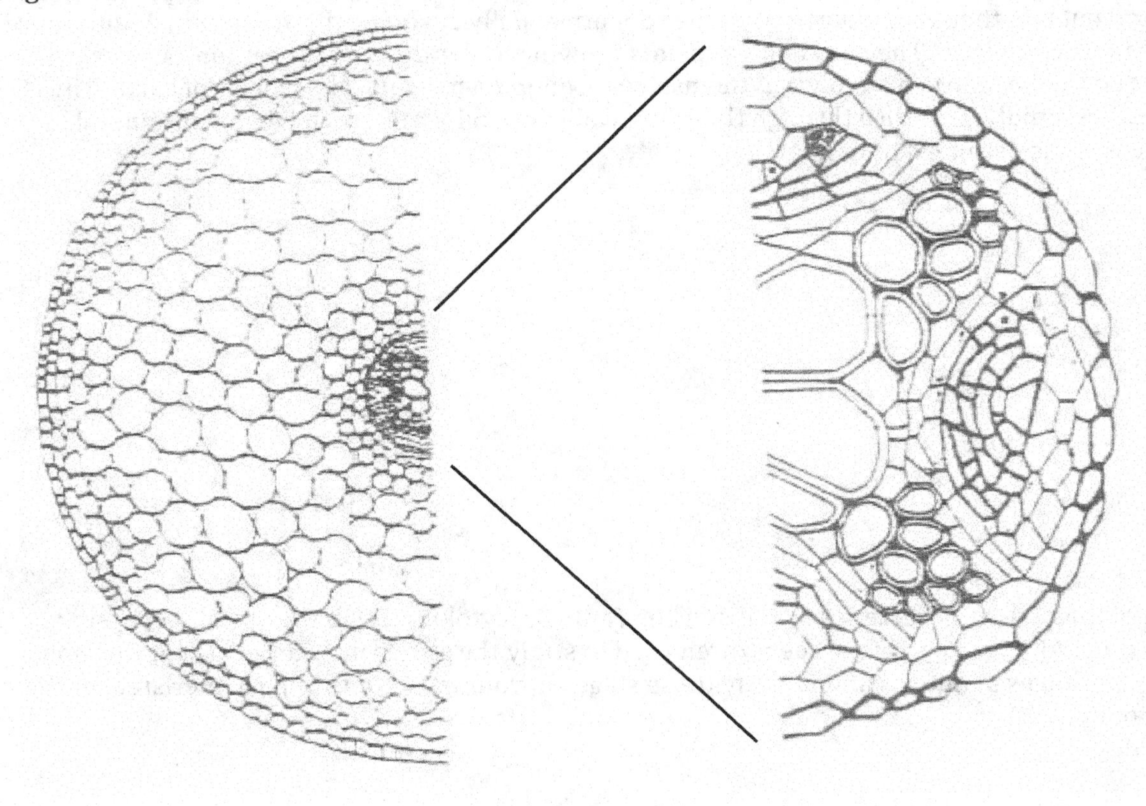

Demonstration:

Examine the prepared slide of a cross-section of a young angiosperm root showing the origin of a **lateral branch** from meristematic activity of the **pericycle**. Note that the lateral root has an apical meristem at its apex! The newly formed lateral root cells elongate and "push" through the root cortex, out into the soil.

Based upon your studies in Laboratories 9-12, complete the following Table.

Table 12.1 General Comparison of Dicots and Monocots

	DICOTS	MONOCOTS
# of Floral Components?		
# of Cotyledons?		
Arrangement of Primary Vascular Bundles in Stem?		
Leaf Venation?		

Based upon your studies in Laboratories 9-12, complete the following Table.

Table 12.2. General Comparison of Stems and Roots

	STEMS	ROOTS
Buds Present/Absent?		
Leaves Present/Absent?		
Nodes & Internodes Present/Absent?		
Pith Generally Present/Absent?		
Origin of Lateral Branches/ Lateral Roots?		

TABLE 12.3: SUMMARY OF SOME CHARACTERISTICS OF TERRESTRIAL PLANTS

Summarize the characteristics of the following plant groups by appropriately completing each "square" of the Table.

GROUP	Vascular Tissue?	True Roots, Stem, Leave?	Homosporous/ Heterosporous?	Antheridia/ Archegonia?	Flagellated/Non-Flagellated Sperm?	Pollen Tube?	Seeds?	Flowers?	Fruits?
"BRYOPHYTES"									
PSILOTOPHYTA									
LYCOPHYTA									
SPHENOPHYTA									
GYMNOSPERMS:									
"PRE-CONIFERS"									
"CONIFERS"									
ANGIOSPERMS									

*If true leaves are present, indicate whether they are microphylls or macrophylls.

SPECIAL EXERCISE ON WOODY PLANT IDENTIFICATION

The purpose of this exercise is to provide you with an opportunity to (1) learn to recognize the major features of some woody plants important in identification of a given specimen, (2) learn, after some experience, to recognize on sight some of the more common trees that occur in Arkansas and (3) learn how to employ the use of a dichotomous key in identification of biological specimens.

We will, as a laboratory activity, utilize one meeting for a campus field trip to introduce you to the use of the key that follows. Since we will be going out regardless of the weather at the time of the laboratory meeting, you may wish to remove this and the following pages from your lab manual, enclosing them in plastic page covers to protect them in the event of rain. All of the terms utilized in the key can be found in your lecture textbook. It would prove helpful to read over the key and develop an understanding of any terms with which you are unfamiliar before the laboratory meeting.

The key that follows is constructed in dichotomous form with characteristics given in sets (couplets) of opposing pairs. Starting with the couplet numbered, "1", one chooses the best fit between that pair of descriptors for the specimen under study, then proceeds to the next couplet indicated by the key number. Similarly, the next pair of choices is considered, et seq., until a generic name is reached at the end of a choice. Sometimes it proves helpful, if not in fact necessary, to "back-track" in the key, following both dichotomies of choice before arriving at a satisfactory determination.

KEY FOR IDENTIFICATION OF SOME COMMON TREE GENERA
(Adapted from Balbach & Bliss, 1991)

1a. Leaves slender, needle- or scale-like, 3 mm or less in width, often evergreen [Figure 1a].......2
1b. Leaves wider, not needle- or scale-like, >3 mm in length, evergreen or deciduous [Figures 1b, 2a,b]........10

2a. Leaves borne in groups ("clusters")........3
2b. Leaves borne singly along the stem........4

3a. Leaves 2-5 in a group ("cluster")........Pinus (Pine)
3b. Leaves >5 in a group ("cluster")........Larix (Larch or Tamarack)

4a. Leaves mostly scale-like to triangular, opposite or whorled........5
4b. Leaves needle-like (sometimes somewhat flattened), alternate or scattered along the stem...6

5a. Smaller branches flattened; cones longer than wide, brown and woody at maturity; leaves scale-like and opposite........Thuja (Arbor Vitae)
5b. Smaller branches not flattened; cones round, bluish and berry-like at maturity; leaves scale- or needle-like, mostly in whorls of 3........Juniperus (Cedar)

6a. Leaves soft and delicate, not >1.5 cm in length, alternate and borne on green branchlets 4-8 cm in length, superficially resembling a compound leaf ....Taxodium (Bald Cypress)
6b. Leaves rigid, solitary and borne on brownish branches........7

7a. Leaves four-sided in cross-section ....Picea (Spruce)
7b. Leaves flattened in cross-section........8

8a. Plant a shrub; leaves without lines of stomata on lower surface........Taxus (Yew)
8b. Plant a tree; leaves with two lines of stomata on lower surface........9

9a. Leaves 8-20 mm in length, with rounded tips and dispersed into two rows along the stem; cones small (<2.5 cm in length)........Tsuga (Hemlock)
9b. Leaves 2-6 mm in length, with pointed tips and rarely, if at all, dispersed into two rows along the stem; cones larger (>3 cm in length)........Abies (Fir)

10a. Leaves fan-shaped, with parallel venation, the veins often dichotomously branched (Figure 1b)........Ginkgo biloba (Maidenhair Tree)
10b. Leaves not fan-shaped, with reticulate venation (Figures 2a, b)........11

11a. Branches (at least the younger ones) bearing spines 12
11b. Branches without spines 15

Leaves wide, not needle- or scale-like, with reticulate venation; branches with spines.

12a. Leaves simple (Figure 3a) 13
12b. Leaves compound (Figures 3a, b)........14

13a. Leaf margins entire (Figure 4a)........Maclura pomifera (Osage Orange)
13b. Leaf margins toothed, sometimes shallowly lobed (Figure 4b)........Crataegus (Hawthorn)

14a. Spines not forked, up to 2 cm in length; leaves once- compound, with 7-17 leaflets............Robinia pseudoacacia (Black Locust)
14b. Spines often forked, up to 7 cm in length; leaves once- or twice-compound, mostly with >18 leaflets............Gleditsia triacanthos (Honey Locust)

15a. Leaves simple............16
15b. Leaves compound............44

16a. Leaves alternately arranged (Figure 5b)............17
16b. Leaves oppositely arranged or in whorls (Figures 5a, c)............42

17a. Leaf margin entire............18
17b. Leaf margin toothed (or appearing to be shallowly lobed) or deeply lobed (Figure 4c)......23

Leaves wide, not needle- or scale-like, with reticulate venation; branches without spines; leaves simple, alternate, with entire margins.

18a. Leaves heart-shaped (cordate)............Cercis canadensis (Red Bud)
18b. Leaves not heart-shaped............19

19a. Mature leaves 20 cm in length, or longer............20
19b. Mature leaves 18 cm in length, or shorter............21

20a. Terminal bud large, enclosed within a single bud scale, the scale usually with gray hairs; stipule scars encircling twigs at nodes............Magnolia (Magnolia)
20b. Terminal bud small, thin, naked, covered with short brown hairs; stipules and stipule scars lacking............Asimina triloba (Pawpaw)

21a. Leaves bristle-tipped, leathery; fruit an acorn............Quercus (Oak)
21b. Leaves not bristle-tipped, not leathery; fruit not an acorn............22

22a. Young twigs smooth; pith of older twigs diaphragmed (Figure 6b); leaf scars with 3 vascular bundle scars............Nyssa sylvatica (Sour Gum)
22b. Young twigs hairy; pith of older twigs chambered (Figure 6a); leaf scars with 1 vascular bundle scar............Diospyros virginiana (Persimmon)

23a. Leaf margin singly- or doubly-toothed or appearing shallowly lobed (Figure 4b)............24
23b. Leaf margin deeply lobed (Figure 4c)............37

Leaves wide, not needle- or scale-like, with reticulate venation; branches without spines; leaves simple, alternate, with toothed margins.

24a. Leaf margin singly-toothed (Figure 4b)............25
24b. Leaf margin doubly-toothed (Figure 4b)............33

25a. Sap milky............Morus (Mulberry)
25b. Sap not milky............26

26a. Leaves triangular or heart-shaped (cordate), about as wide as long............27
26b. Leaves longer than wide, the greatest width at the midregion............28

27a. Leaves asymmetrically shaped with an oblique base, with several (usually 3) large veins (Figure 2b) arising from the petiole .... Tilia (Basswood)
27b. Leaves symmetrically shaped, pinnately veined (Figure 2a), with a prominent midrib extending from petiole to apex .... Populus (Cottonwood)

28a. Teeth of leaf margin coarse (2 or </cm) or leaf margin undulate; buds mostly 15-25 mm in length, pointed .... Fagus (Beech)
28b. Teeth of leaf margin fine (3 or >/cm); buds usually <15 mm in length 29

29a. Petiole with 1-2 small glands near the base of the leaf blade .... Prunus (Plum; Peach; Cherry)
29b. Petiole without glands 30

30a. Leaves at least 3 times longer than wide .... Salix (Willow)
30b. Leaves 2 times, or less, longer than wide 31

31a. Leaves asymmetrically shaped, with 3-5 large veins arising from the petiole; pith of older twigs chambered (Figure 6a) .... Celtis (Hackberry)
31b. Leaves symmetrically shaped, with one main midrib; pith of older twigs not chambered (Figure 6c) 32

32a. Buds elongated, with many bud scales; bark of trunk scaling off in narrow vertical strips .... Ostrya virginiana (Hop Hornbeam)
32b. Buds ovoid, with few bud scales; bark of trunk not scaling off in narrow vertical strips .(Apple)

33a. Leaf blade asymmetrical at the base .... (Elm)
33b. Leaf blade symmetrical at the base 34

34a. Petiole with 1-4 small glands near the base of the leaf blade .... Prunus (Plum; Cherry)
34b. Petiole without glands 35

35a. Bark of trunk and larger branches silvery- to yellowish-white, peeling horizontally into thin, papery strips .... Betula (Birch)
35b. Bark brown, not peeling into horizontal strips 36

36a. Buds on short stalks, with 2 bud scales .... (Alder)
36b. Buds sessile, not on stalks, with several bud scales .... Ostrya virginiana (Hop Hornbeam)

Leaves wide, not needle- or scale-like, with reticulate venation; branches without spines; leaves simple, alternate, with deeply lobed margins.

37a. Leaves palmately veined (Figure 2b) 38
37b. Leaves pinnately veined (Figure 2a) 40

38a. Sap milky; some of the leaves toothed but not lobed (Figure 4b) .... (Mulberry)
38b. Sap not milky; all of the leaves lobed (Figure 4c) 39

39a. Leaves star-shaped, with 5-7 wedge-shaped lobes, the margins evenly toothed; terminal bud present .... Liquidambar styraciflua (Sweet Gum)
39a. Leaves not star-shaped, 3-5-lobed, the margins not evenly toothed; terminal bud absent .... Platanus occidentalis (Sycamore)

40a. Young branches yellowish-green; leaves and bark aromatic . . . ..Sassafras albidum (Sassafras)
40b. Young branches gray or grayish-brown; leaves and bark not aromatic. 41

41a. Leaf broadly notched at apex, 4-lobed, about as wide as long . . . .
.... Liriodendron tulipifera (Tulip Popular)
41b. Leaf with 5, or more, marginal lobes, longer than wide . . . . Quercus (Oak)

Leaves wide, not needle- or scale-like, with reticulate venation; branches without spines; leaves simple, opposite or whorled.

42a. Leaves 3, less commonly 2, at a node (Figure 5c), ovate to ovate-oblong, with a long, pointed tip, lower surface hairy. . . . . Catalpa (Catalpa)
42b. Leaves opposite (Figure 5a), only slightly, if at all, hairy 43

43a. Leaf margin entire, pinnately veined (Figures 2a, 4a) . . . . .Cornus (Dogwood)
43b. Leaf lobed, palmately veined (Figures 4c, 2b). . . . .(Maple)

44a. Leaves alternate (Figure 5b) 45
44b. Leaves opposite (Figure 5a) 53

Leaves wide, not needle- or scale-like, with reticulate venation; branches without spines; leaves compound, alternate.

45a. Leaves twice compound 46
45b. Leaves once compound 48

46a. Base of leaflets asymmetrically shaped; midrib of leaflets off-center. . . . . . . Albizzia (Mimosa)
46b. Base of leaflets symmetrically shaped; midrid in center of leaflets 47

47a. Leaflets ovate, the margins entire, with pointed apex . . . . .Gymnocladus dioica (Kentucky Coffee Tree)
47b. Leaflets oval or lanceolate, the margins finely toothed, with rounded apex . . . . . Gleditsia triacanthos (Honey Locust)

48a. Pith of older twigs chambered (Figure 6a); leaflets 11-33 . . . . Juglans (Walnut)
48b. Pith of older twigs not chambered (Figure 6b) 49

49a. Leaves 30-90 cm in length, with 13-41 leaflets, producing an unpleasant odor when crushed
.... Ailanthus altissima (Tree of Heaven)
49b. Leaves usually <30 cm in length, without an unpleasant odor 50

50a. Leaflets <5 cm in length, with rounded or emarginated apex, the margin entire or inconspicuously toothed 51
50b. Leaflets >5 cm in length, with pointed apex, the margin conspicuously toothed 52

51a. Leaves with a spiny stipule, with 7-17 leaflets . . . . Robinia pseudoacacia (Black Locust)
51b. Leaves without a stipule, typically with 18 leaflets . . . . . . Gleditsia triacanthos (Honey Locust)

52a. Plant a small tree or shrub, with milky sap; buds without bud scales . . . . (Sumac)
52b. Plant a large tree, without milky sap; buds with bud scales . . . . (Hickory)

Leaves wide, not needle- or scale-like, with reticulate venation; branches without spines; leaves compound, opposite.

53a. Leaves pinnately compound (Figure 3b) 54
53b. Leaves palmately compound (Figure 3c) 57

54a. Leaflets 3-5, coarsely and irregularly toothed and lobed (Figures 4b, c) ...... Acer negundo (Box Elder)
54b. Leaflets 5-11, entire or finely toothed (Figures 4a, b) 55

55a. Young twigs 4-angled ....Fraxinus quadrangulata (Blue Ash)
55b. Young twigs rounded, not 4-angled 56

56a. Top of leaf scars concave; petioles hairless or nearly so .... Fraxinus americana (White Ash)
56b. Top of leaf scars straight or nearly so; petioles with velvety hair .... Fraxinus pennsylvanica (Red Ash)

57a. Leaflets usually 7, coarsely doubly-toothed ....Aesculus hipposcastanum (Horse Chestnut)
57b. Leaflets usually 5, finely singly-toothed 58

58a. Plant a small tree, up to 15 m tall; fruit spiny .... Aesculus glabra (Ohio Buckeye)
58b. Plant a large tree, up to 25 m tall; fruit smooth ....Aesculus octandra (Sweet Buckeye)

SPECIAL IDENTIFICATION KEYS FOR LEAF VENATION, COMPLEXITY, MARGINS, ARRANGEMENT, AND THE PITH COMPLEXITY

Figure 1. Leaves Slender, Needle- or Scale-like or Fan-shaped with Parallel Venation

1a. Slender, needle- or scale-like

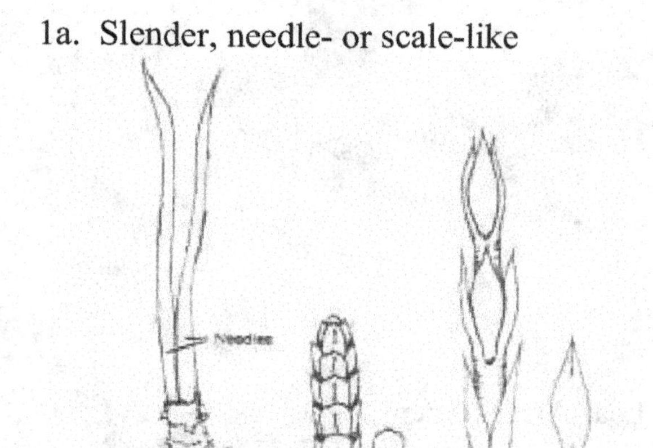

1b. Fan-shaped with parallel venation

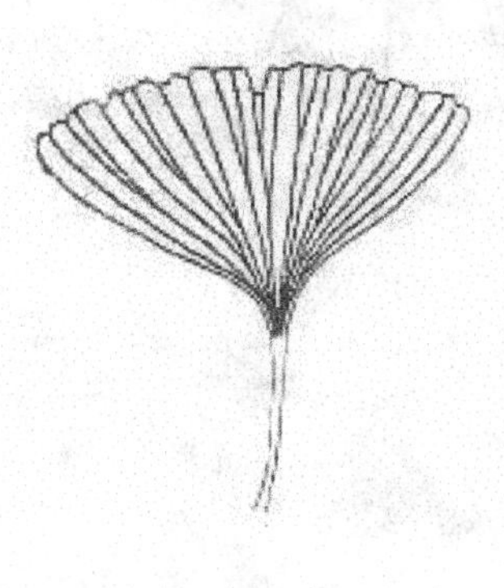

Figure 2. Leaf with Reticulate (Net) Venation

2a. Pinnate venation

2b. Palmate venation

Figure 3. Simple and Compound Leaf Complexities

3a. Simple

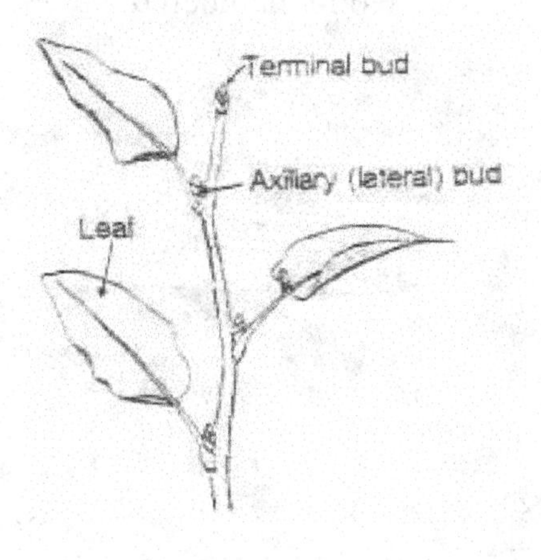

3b. Pinnately Compound

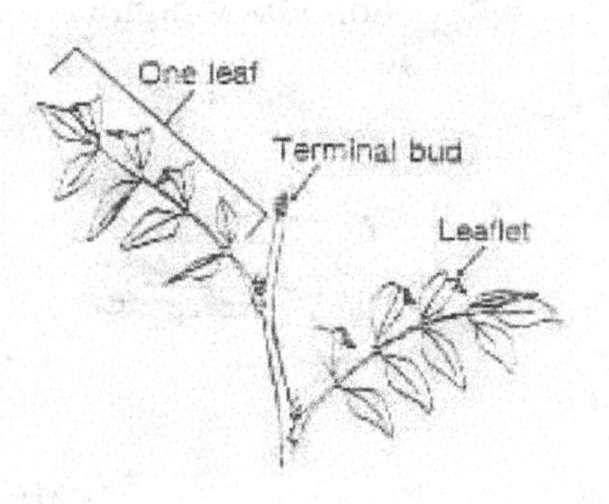

3c. Palmately Compound

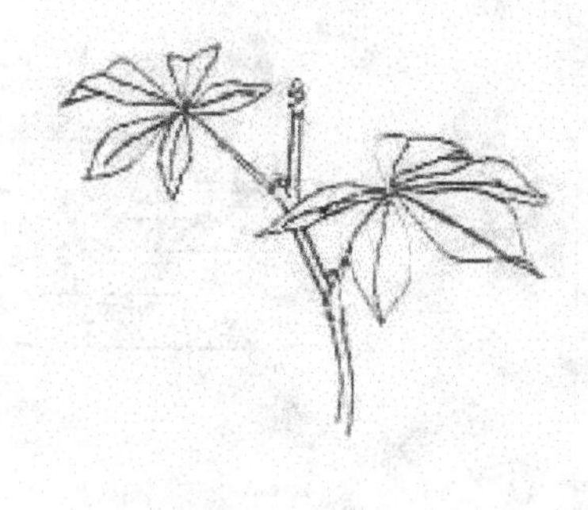

Figure 4. Leaf Margins

4a. Entire 4b. Toothed 4c. Lobed

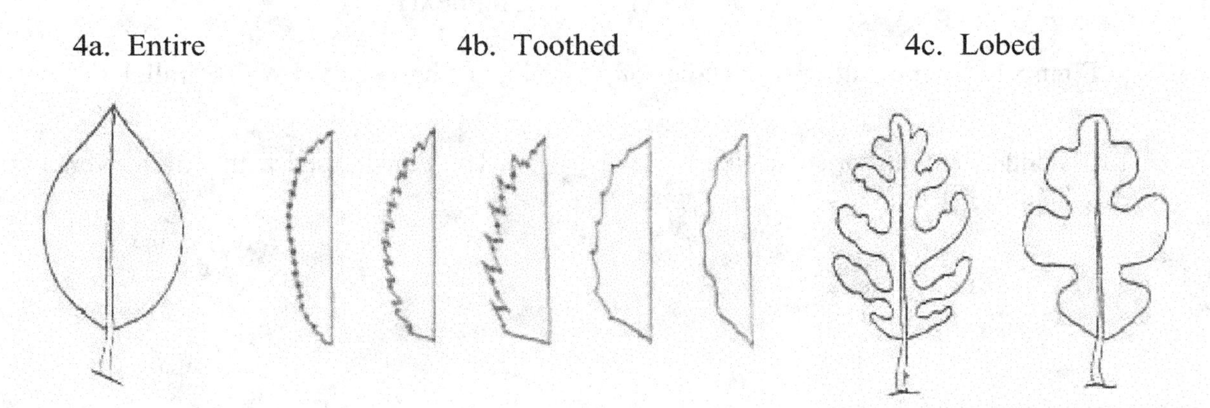

Figure 5. Leaf Arrangement

5a. Opposite 5b. Alternate 5c. Whorled

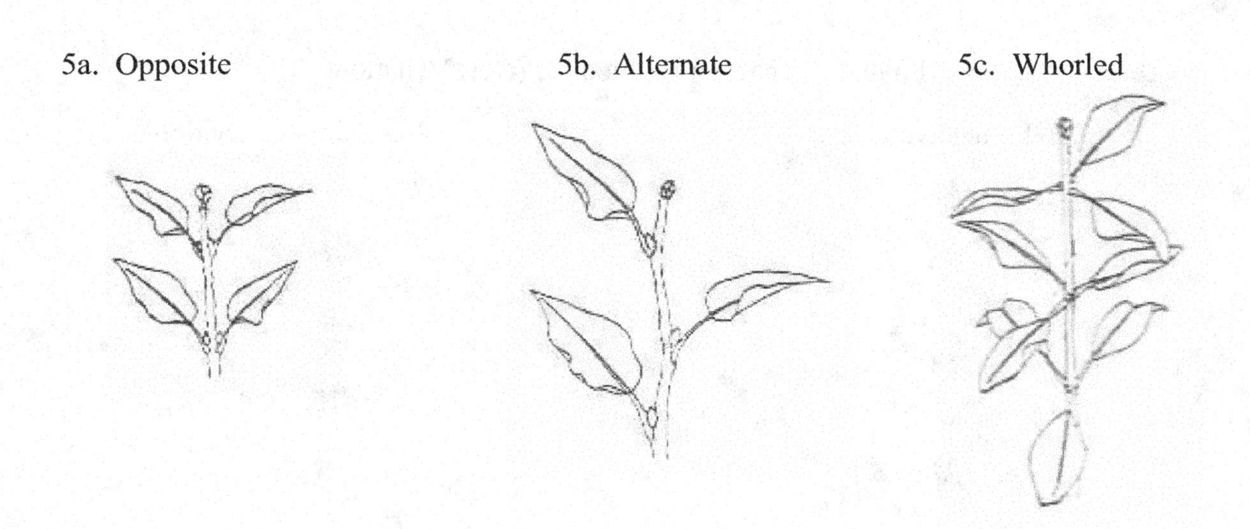

Figure 6. Chambered, Diaphragmed and Non-Chambered Pith

6a. Chambered 6b. Diaphragmed 6c. Non-Chambered

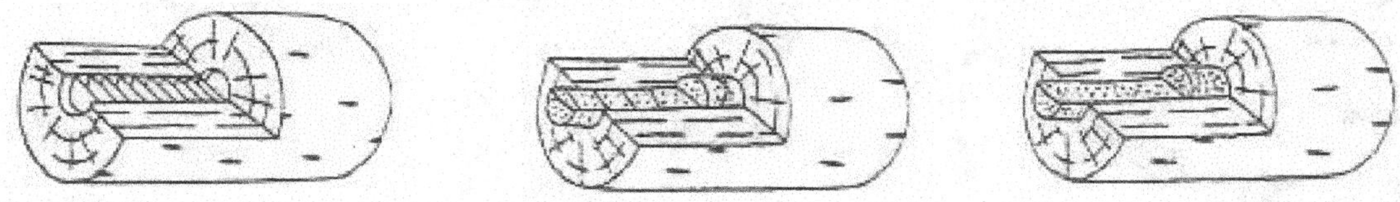

FRUIT DICHOTOMOUS KEY

1A. Fruits fleshy – **go to 2**
1B. Fruits dry – **go to 8**

2A. Fruit simple (true), derived from a flower with 1 pistil – **go to 3**
2B. Fruit compound (false), derived from more than 1 pistil – **go to 7**

3A. 1 seed enclosed in bony endocarp (pit) – **drupe** - peach, cherry, plum, olive, coconut
3B. more than 1 seed – **go to 4**

4A. Leathery or papery endocarp, inferior ovary – **pome** - apple, pear
4B. No bony, leathery or papery endocarp – **go to 5**

5A. Ovary inferior, fruit with rind (gourd family) – **pepo** - cucumber, watermelon, pumpkin, squash
5B. Ovary superior – **go to 6**

6A. Thin skin - **true berries** - tomato, grape, coffee
6B. Leathery skin with oils (citrus family) – **hesperidium** - all citrus fruits

7A. Fruit derived from 1 flower with more than 1 pistil – **aggregates** - strawberry, rose hip, raspberry
7B. Fruit derived from more than 1 flower - **multiple fruits** - pineapple, sweetgum, fig, mulberry

8A. Indehiscent – **go to 9**
8B. Dehiscent – **go to 12**

9A. Fruits with a wing - **samara** - ash, maple, tulip tree
9A. Fruits without a wing – **go to 10**

10A. With a hard shell – **nut** - acorn, macadamia
10B. Without a hard shell – **go to 11**

11A. Pericarp fused entirely to seed coat (grass family) – **caryopsis** (grain) - all cereals
11B. Pericarp not fused entirely to seed coat – **achene** - sunflower

12A. Derived from several fused carpels, opening by slits, pores or cap – **capsule** - cotton, poppy
12B. Derived from 1 or 2 fused carpels, dehiscing lengthwise – **go to 13**

13A. Fruit with persistent septum (replum) (mustard family) – **go to 14**
13B. No persistent septum – **go to 15**

14A. Long & thin – **silique** - mustard
14B. Short & fat – **silicle** - shepherd's purse

15A. Dehiscent along 1 edge – **follicle** - milkweed, magnolia
15B. Dehiscent along 2 edges (legume family) – **legume** - peanuts, all beans